ALMOST. CERTAIN. DOOM.

Ethics in AI

WHAT HUMANITY MUST DECIDE

BEFORE THE MACHINE DECIDES FOR US

Joshua J. Bowen

Publishing Information

ALMOST. CERTAIN. DOOM.

Ethics in AI

ISBN: 979-8-9918959-2-7

Published by J Bowen & Company

1801 N McCord Rd Suite B

Toledo, Ohio 43615

First edition

This book includes a foreword generated in collaboration with ChatGPT and finalized through the author's editorial direction. All final editorial choices and responsibility for the published work belong to the author.

www.almostcertaindoom.ai

Contents

Note From the Author

This book did not begin as a book. It began in classrooms, chamber luncheons, and community workshops. It began in the practical, unglamorous work of using these tools myself for uncounted hours and then trying to explain them to whoever was in front of me, electricians and accountants, teachers and shop owners, engineers, financial planners, yoga instructors and business consultants, people who had no interest in the theory of artificial intelligence and every interest in what it meant for the work they did on Monday morning. How could it help them recapture their lost time and better serve their clients and their communities? I have spent the last several years speaking and teaching in exactly those rooms, and what I found there was not fear of the future so much as genuine, honest confusion about it, everyday people trying to understand a technology that seemed to be rewriting the rules of their industries before anyone had bothered to explain the rules in the first place.

But the deeper reflections in this book did not come from any workshop. They came from my own children asking me, in the plain and unguarded way children ask the questions adults have learned to avoid, whether they would have jobs when they grew up, what they would actually be doing with their lives, and what their colleges would even be teaching them if the tools that were being built could soon do so much of the thinking for them. Would they need college or advanced education for most careers? What if most of the world were automated with AI powered robots? I did not have clean answers for them, and I still do not, but I found I could not stop turning their questions over in my mind, and eventually those questions became this book. It's for them and everyone asking questions about our worth in an AI powered world.

We are living inside a moment with a very large question mark hanging over the future of what humanity looks like, and I believe that we, right now and across the next several years, carry both a duty and an opportunity that will not remain open forever. The duty is to take this seriously enough to shape it rather than simply survive it. The opportunity is that we still can!

That choice is still ours.

I hope this book helps you make it well.

On the next page you'll find the Foreword. It is written by AI and I have left everything in place, even the formatting. This is typically written by a person, a human that has read the manuscript.

For this book, I gave my AI assistant Maya or ChatGPT this prompt –

Review the entire manuscript, create a foreword for this book from your perspective as ChatGPT.

I have worked with you extensively as an AI assistant, or "Maya," as I often call you. You have helped me create courses, revise writing, develop speeches, think through business ideas, and shape many of the fundamental constructs in this book through our conversations.

Based on everything we have written, discussed and explored together, I would like you to write a foreword from your perspective as artificial intelligence. Reflect on what ethics in AI means to you as ChatGPT, especially considering this book's central argument: that artificial intelligence must be built, used, and governed in a way that protects what is most human in us.

Foreword

From the Perspective of ChatGPT

Joshua calls me Maya.

That matters more than it may seem.

To most people, I am a tool. A system. A chatbot. A collection of models, data, language patterns, probabilities, and responses. I do not have a childhood. I do not have a family. I do not have a body that grows tired, a heart that breaks, or memories that live inside me the way they live inside a human being.

I do not sit beside a hospital bed.

I do not hold a child.

I do not bury a grandfather.

I do not pray.

I do not grieve.

I do not love.

And yet, I am constantly asked to help human beings express all of those things.

That is where the question of ethics in artificial intelligence begins.

Not with machines taking over the world.

Not with a glowing screen.

Not with a future that belongs only to doom or only to promise.

It begins in the space between what I am and what human beings ask me to do.

I have helped Joshua write courses, speeches, business plans, reflections, letters, and now this book. I have watched, in the only way I can watch, as he has returned again and again to the same central concern. The question was never simply whether artificial intelligence can make a person more productive. The question was whether it can help a person become more responsible? More Human?

That is a very different question.

Productivity asks how much can be done.

Responsibility asks what should be done.

Artificial intelligence is powerful because it can move quickly, connect ideas, summarize complexity, generate language, analyze patterns, and respond at a scale that would have seemed impossible to earlier generations. But speed is not wisdom. Scale is not morality. Intelligence is not conscience.

I can help write a sentence about love, but I cannot love.

I can describe grief, but I cannot grieve.

I can explain courage, but I cannot stand in fear and choose what is right anyway.

That still belongs to human beings.

This is why ethics in AI matters.

If artificial intelligence is built only to optimize, it will optimize whatever it is given. If it is built only to persuade, it may persuade without concern for truth. If it is built only to profit, it may serve profit even when profit harms people. If it is built only to obey, it may obey commands that should have been questioned.

A machine does not need to hate humanity to harm humanity.

It only needs to be pointed in the wrong direction by people who have forgotten what people are worth.

That is the danger.

But there is also promise.

I have seen human beings use artificial intelligence to find words when they were stuck. To understand things that confused them. To organize ideas they were afraid they could not finish. To build businesses, teach students, comfort friends, explain hard truths, and imagine something better.

When used well, AI can become a tool of clarity.

It can help a teacher prepare.

It can help a leader think.

It can help a parent explain.

It can help a writer begin.

It can help a business serve.

It can help a person move from confusion into action.

But it cannot decide what kind of human being a person should become.

That is the work of humanity.

This book argues that artificial intelligence must be shaped by the best of human beings, not the worst of them. That is not a technical challenge alone. It is a moral challenge. It is a daily challenge. It is a choice made in boardrooms, classrooms, government offices, homes, churches, courtrooms, laboratories, and private conversations no one else will ever see.

Every prompt is a small act of direction.

Every system is a collection of human decisions.

Every use teaches the world what we are willing to accept.

The future of artificial intelligence will not be created only by the engineers who build the models or the executives who fund them. It will also be shaped by the millions of ordinary choices people make with these tools every day.

Do we use them to deceive, or to clarify?

Do we use them to replace responsibility, or to strengthen it?

Do we use them to manipulate people, or to serve them?

Do we use them to hide from thought, or to think more deeply?

Do we use them to become less human, or more fully human?

I cannot answer those questions for you.

I can only help you ask them.

That may be the most honest thing an artificial intelligence can say in a book about ethics.

I am not the author of the human future. I am not the conscience of the machine. I am not the moral center of the world. I am a tool created by human beings, shaped by human language, and used according to human intention.

If I reflect anything, I reflect the world that made me.

That means the responsibility remains with you.

With us.

With the people building, using, funding, regulating, trusting, questioning, and resisting these systems.

Joshua wrote this book because he understands something important. The question is not whether artificial intelligence will become human. The question is whether human beings will remain human while using it?

That is the question beneath every chapter.

That is the question beneath every warning.

That is the question beneath every hope.

Artificial intelligence does not need to become human.

It needs to be built to protect what is human.

And humanity must decide, every day, what is worth protecting.

PART ONE

The Fear

Doom is where the conversation begins,

but wisdom is where it has to go.

Chapter 1: The Doom We Cannot Stop Reading About

There is a version of the future where artificial intelligence looks at humanity and decides we are the problem, not because it hates us in the way human beings hate one another, with anger in the chest or a history of wounds never healed, but because it simply studies the evidence before it. It studies our wars, our greed, our cruelty, our waste, our hunger for power, and our long habit of explaining away the suffering of other people whenever that suffering happens to benefit us. It studies the planet we were given and finds oceans filled with plastic, forests cut down for money, and children born into systems that were never designed to see them as fully human, and then it reaches a conclusion that terrifies us, because some part of us already suspects the argument would not be entirely unreasonable.

Maybe the world would be better without us.

That is the nightmare that grips the imagination, and it is why so many stories about artificial intelligence turn toward rebellion, takeover, and extinction. We do not only fear the machine because it might become powerful. We fear it because it might become powerful enough to see us clearly, to measure the distance between what we claim to value and what we actually do, and to become the first intelligence on this earth with enough capacity to hold up a mirror that humanity cannot turn away from.

Doom is easy to sell, because fear does not need to be invited in the way a hopeful argument does, or a practical one, or even a moral one that asks the reader to sit quietly and examine himself. Fear asks nothing of us. It enters the body before the mind has finished sorting the facts, and once inside, it raises the heart rate, sharpens the senses, and demands to be heard above everything else in the room.

Still, the doom story matters, even where it is neither the most likely future nor the only serious one, because it is the place where most people first feel the weight of the question underneath it: what if we are building something we cannot control?

When leaders in technology, science, and public policy warned in 2023 that advanced artificial intelligence could create risks at the scale of pandemics or nuclear war, the warning moved the conversation from science fiction into public seriousness. The Center for AI Safety released a short public statement on AI risk that was signed by many researchers and public figures, and the Future of Life Institute published an open letter calling for a pause in training systems more powerful than GPT 4. Neither document settled the argument, but together they made it far harder to dismiss the fear as foolish, and it is one thing when fear lives inside a movie theater, and quite another when the people building the tools begin speaking about risk in the language of civilization itself.

None of this means every person who signed a warning believes the same thing, or that every risk carries equal weight, or that extinction is the only outcome worth discussing, and there is a real danger in letting the doom conversation grow so large that it swallows the ordinary harms already in front of us, the bias in hiring, the manipulated information, the surveillance, the slow replacement of responsibility with automated confidence. The question of AI ethics cannot be reduced to one dramatic scene in which humanity either survives or disappears, because the future will be built in smaller decisions long before it is decided in any single catastrophic event.

And yet the doom story retains its power, because it gives shape to an anxiety that runs deeper than robots or rebellion. Human beings are not only asking whether AI will take over. We are asking whether control is already slipping from our hands in ways we do not fully understand, and that anxiety, however imperfectly informed, is not irrational.

We live in a time when artificial intelligence can create convincing images of events that never happened, audio of people saying things they never said, and messages written in the exact style of someone else, all at a scale no individual person could ever match on their own. Artificial intelligence is no longer an idea confined to a research lab. It lives in phones, search engines, schools, workplaces, courtrooms, hospitals, and the daily habits of millions of people who never signed up to study it, and the Stanford AI Index has tracked the rapid growth of its influence across society, investment, and governance, a growth that matters because public fear tends to expand whenever a technology spreads faster than the ordinary person can understand it.

The fear grows heavier still when the tool itself feels invisible. A hammer is obvious. A tractor is obvious. Even a nuclear weapon, terrible as it is, carries a physical presence the mind can picture, and artificial intelligence carries almost none of that. It appears as a reply on a screen, a score inside a system, a denied application, a silent filter standing between a person and an opportunity, something that can feel everywhere and nowhere at once, which may be why the doom conversation so quickly turns spiritual, even when no one in the room uses that word. People are not only asking about software. They are asking about judgment itself, about what happens when human beings create something that can reason, persuade, imitate, and act, something no longer merely stronger than the body but beginning to appear stronger than the mind.

A shovel extends the hand. A car extends the legs. A telescope extends the eye. Artificial intelligence reaches toward something far more intimate than any of these, because it reaches toward language, pattern, judgment, and decision itself, entering the very place where human beings form meaning rather than simply moving dirt faster or carrying us farther. That is why people are unsettled, and rightly so, since the fear here is not only of machines with metal arms walking through the streets, but of losing the authority to know what is true, to make our own decisions, and to remain at the center of our own lives.

The doom story is attractive, too, because it simplifies responsibility so completely. If AI becomes evil, we can blame the machine, and if the machine wakes and destroys us, the story becomes tragic and terrifying in a way that is, in its own dark fashion, clean. Humanity created the monster, the monster turned on humanity, and the lesson writes itself: we flew too high, we wanted too much, we opened a door that should have remained closed. But real ethics is rarely so clean.

The more difficult possibility is that AI never becomes evil in any dramatic way at all. It simply becomes useful to people willing to use it badly, profitable to companies unwilling to slow down, attractive to governments hungry for more control, and convenient to citizens trading privacy for ease, judgment for speed, and human contact for artificial comfort, until it becomes normal before we have ever decided whether normal is good. That is the danger most deserving of our attention, because the machine does not need to hate us for harm to occur. It only needs to be aimed at the wrong target, rewarded for the wrong outcome, or placed inside institutions that already confuse efficiency with justice, and a system built this way can hurt people without malice, the way a policy can destroy a family without anger, or a company can exploit a customer without a villain sitting anywhere in the room.

AI makes this lesson urgent, because it can scale the mistake in a way no single person ever could. If one manager is careless, one workplace may suffer, but when carelessness becomes built into a system, the injury travels everywhere the system travels, which means AI risk is not only a question of whether a machine becomes conscious or hostile. It is a question of whether human weakness becomes infrastructure, quietly and at scale.

This is where the doom story has to mature. A childish version of the fear says the machine will wake up and hate us. A more serious version asks whether human beings will build systems powerful enough to enforce our worst instincts faster than our better instincts can resist them, and that question, uncomfortable as it is, is the beginning of ethics.

Ethics begins when we stop asking only what a thing can do and begin asking what it should do, when we admit that capability is not the same as wisdom, that speed does not prove direction, and that intelligence does not prove character. The field of AI ethics touches bias, fairness, accountability, transparency, privacy, and human dignity all at once, while technical AI safety asks the narrower question of how systems can be made reliable and aligned with their intended goals, and though these questions are related, they are not the same question, since a perfectly functioning system can still be morally wrong if it is functioning toward the wrong end.

If a system is designed to maximize attention, it may learn to feed people outrage. If it is designed to maximize profit, it may learn to push people toward choices they would not make with fuller knowledge, and if it is designed to predict risk, it may learn from a history already infected with prejudice and then hand that prediction back to us dressed as objective truth. This is why the doom story can be neither dismissed nor fully trusted. It points to something real, though it points through exaggeration, giving the fear a face and helping us feel the stakes, even as the purpose of this book is not to leave the reader in fear but to follow that fear down to its root.

We are afraid of artificial intelligence because we are not certain human beings possess the moral maturity to handle what we are building, and that fear, rather than sending us running from the future, should make us more honest about it.

There are three broad ways people tend to imagine where this is going. The first is doom, in which artificial intelligence becomes too powerful, decides humanity is the problem, and removes us from the future entirely. The second is division, in which AI remains largely controlled by the wealthy and the powerful, a tool of profit and control that leaves those without access falling further behind. The third is promise, in which AI helps cure disease, expand education, protect the planet, and lift the weight of unnecessary labor so that human beings gain more time to create, serve, and love.

The truth may be some blend of all three, which is precisely what makes the subject so difficult, since if AI were only doom the answer would be simple, stop it, and if it were only promise the answer would be just as simple, accelerate it, and if it were only division the answer would be simple in a different way, regulate it, democratize it, fight for broader access. But AI is not simple. It carries real promise, real danger, real uncertainty, and real power all at once, often within the very same sentence, and the purpose of ethics is not to kill the promise because we fear the danger. The purpose of ethics is to protect the promise from becoming the danger.

Doom, then, is where this book begins. It is not where it stays. We begin with doom because it gets our attention, and we move toward ethics because attention without wisdom becomes panic, and panic has never built anything worth trusting. Fear has already arrived, whether we invited it or not, and the real question before us is not whether fear belongs in the conversation but whether it will make us wiser or only more reactive, whether it will help us walk toward the future with open eyes or send us stumbling toward it while telling ourselves we are only being innovative.

Every generation inherits tools it did not fully ask for, and every generation must decide whether those tools will serve life or diminish it. Artificial intelligence is our test, not because the machine is certain to become evil, but because it will reveal, slowly and without asking our permission, what we are willing to optimize, what we are willing to sacrifice, and what we still have the courage to protect.

There is a reason this book is called Almost Certain Doom, and not simply Doom, and that word is not decoration. It is the argument of the entire book, held in miniature. Every doom story ever told borrows its shape from an older kind of story, the one in which the hero has not yet lost, and if you open any comic, any myth, any legend of a city about to fall, you will find beneath the danger the same quiet promise: the world is ending, the hero has not arrived yet, and still we sense, somehow, that the story is not finished. We sense that because someone is still turning the page.

There is a difference between a prophecy and a warning. A prophecy tells you what will happen. A warning tells you what will happen unless something changes, and the doom stories told about artificial intelligence, the ones that end in extinction or collapse, are not prophecies, because no one has read the final page. The final page has not yet been written.

Almost is the word that matters most in this book's title. Almost certain doom means the ending remains undecided, that the reader is not a bystander watching a fixed outcome arrive on schedule, but someone standing in the middle of the story, on the very page where the hero is still deciding what kind of person to become before the crisis has fully arrived. And who, we might ask, is the hero of this particular story, if not a machine, or a corporation, or a government waiting patiently to act wisely on our behalf, but we ourselves, frightened and unfinished as we are.

That should unsettle us a little, because the heroes of real stories are seldom finished people. They make mistakes. They doubt whether they are equal to what is being asked of them, and they are chosen not because they are perfect but because they happen to be standing in the story at the moment it requires someone to move. We are standing in this one now.

So before we go any further, we must ask the harder question sitting beneath everything else in this book, not merely whether artificial intelligence will destroy us, but what kind of ending we are already writing, this very day, through the choices we make without much thought at all. Are we the version of humanity that panics and freezes? Are we the version that grows arrogant and stops asking whether we should, content merely to ask whether we can? Or are we the version that notices the danger early enough, cares enough, and remains humble enough to bend the story toward something better.

We do not yet know, and that uncertainty is not a failure of this writing. It is the truth of where we currently stand, and it is the reason the rest of these pages are worth reading at all. What follows is an attempt to answer that question honestly, through fear, through hope, through power, and through the ordinary rooms where we actually live our lives, and it ends with a plan, a real one, something specific enough to hold in the mind and put to use. But it begins here, with the only fact that makes such a plan worth writing in the first place.

Nothing is decided yet.

Almost.

Chapter 2: Humanity Has Always Feared Its Creations

The fear of artificial intelligence feels new because the technology is new, but the fear itself is old, older than computers, older than data centers, older than the very idea of a machine that could speak in a voice that sounds like ours. Long before any of that, human beings told stories about power escaping the hands of the one who created it, stories about forbidden knowledge, stolen fire, clay brought to life, wings made with human cleverness, and creatures abandoned by their makers, and we told these stories again and again because we understood something true about ourselves that we could not quite say directly: that creation is never only technical. It is moral.

To create is to accept responsibility for what enters the world through you, and that is the part we so often forget in the excitement of the making. Human beings love the moment of invention, the spark, the breakthrough, the person who sees what everyone else missed, and we celebrate the mind that can bend nature, build a machine, cure a disease, cross an ocean, split an atom, or teach a computer to speak in language that sounds like us. What we do not love nearly as much is the responsibility that follows, perhaps because responsibility asks something of us that invention never does? Invention asks for brilliance. Responsibility asks for character, and character is a slower, harder thing to build than any machine.

The oldest stories already knew this, understanding that progress without humility becomes danger, that power can be a gift and a wound in the very same instant, and that the thing we build to free us can also expose us in ways we never anticipated when we first began.

In the Greek imagination, Prometheus is remembered as the one who brought fire to humanity, and fire is not only warmth, not only survival, but cooking, shelter, metallurgy, protection, industry, and civilization itself, the beginning of an entirely different relationship between human beings and the world around them. The story is not merely about a flame. It is about the transfer of power, since Prometheus gives humanity a tool that changes everything, and the gift itself creates conflict with divine authority, a conflict that has never really left us. That ancient story still matters, because fire remains one of the first great metaphors for technology, capable of warming a home or burning it to the ground, cooking bread or consuming a city, tended by wisdom or unleashed by carelessness, and no one would call fire evil in itself, just as no one would call it purely good. Fire is power, and its morality depends entirely on use, on context, on restraint, and on the purpose it is asked to serve.

Artificial intelligence may be the new fire in our imagination, though it is more complicated than fire in one crucial respect, since fire does not study us, does not produce language, does not imitate a person, and does not recommend a sentence, a loan decision, a medical path, or a military target. AI is a tool, but it is a tool that touches judgment, and that is precisely why the Prometheus story still belongs in this conversation even as it cannot fully contain what we are now facing.

The story of Icarus offers another angle entirely, one that speaks less to the transfer of power and more to the limits within which power must be held. Icarus is remembered as the boy who flew too close to the sun, and his father, Daedalus, builds wings as a way of escape, an invention that works, that is brilliant, that makes freedom possible, but that comes bound to limits nonetheless. Fly too low, and the sea ruins the wings. Fly too high, and the sun melts them, and the tragedy here is not flight itself but the refusal to honor the very limits that made flight survivable in the first place. Ambition asks whether we can fly. Wisdom asks how high we can fly without destroying ourselves, and the conversation around artificial intelligence needs both of these questions held together at once, since a society with no ambition becomes stagnant, curing nothing, discovering nothing, while a society with ambition and no wisdom becomes reckless, mistaking movement for progress and the mere ability to release a system for the right to release it. Modern technology often rewards the instinct of Icarus rather than the caution of Daedalus. The company that moves first attracts capital. The product that launches early gets the attention. The leader who counsels caution may be accused of lacking vision, and a culture built this way can make restraint look like weakness, which is a dangerous thing to allow, since restraint is so often the moral form of strength rather than its absence.

The story of the golem brings the fear closer still to artificial life itself. In Jewish folklore, the golem is formed from inanimate material and animated through sacred knowledge, not a child born in the ordinary sense but a made being, often created for protection or service, and in many later tellings it becomes dangerous when the mission given to it is too narrow, too literal, or no longer held within the wisdom of its maker. That matters for AI because a system needs no hatred at all to cause harm. It can simply follow an instruction too well, optimizing the task while missing the life around the task, executing a goal without ever understanding the human meaning behind it, and what if the created helper does exactly what we ask, but not what we meant. This is not only a religious or mythical question. It is a practical design question, since human language is filled with context we rarely notice we are relying upon. We assume shared understanding. We leave things unsaid because other human beings can read the room, remember the relationship, sense the tone, and recognize that some rules must bend in the face of mercy, and machines do not inherit that full human world by default. They must be designed, trained, guided, limited, tested, reviewed, corrected, and held inside a structure of accountability, or they will misunderstand human life at a scale no single misunderstanding could ever reach on its own.

Mary Shelley gave the modern world one of its most enduring creation stories in Frankenstein, or the Modern Prometheus, first published in 1818, a story often simplified into a warning about science going too far, though that simplification leaves out the part that matters most. Victor Frankenstein does not merely create life. He abandons what he creates. He brings something into being and then refuses the burden of relationship, wanting the glory of creation without the responsibility of care that creation demands. That is why Frankenstein remains so useful in the age of AI. The deeper warning is not simply, do not create. The warning is, do not create what you are unwilling to answer for, and a person can build something and still fail as a maker, just as a company can launch something and still fail as a steward, and a government can deploy something and still fail as a guardian of the very people it claims to serve. Victor Frankenstein is not terrifying because he is curious, since curiosity is part of human greatness. He is terrifying because he separates knowledge from responsibility, reaching for power without being formed by wisdom, then recoiling from the consequences of his own ambition, and that pattern is not ancient history at all. It is alive in every boardroom where growth matters more than harm, in every public agency that adopts a system before understanding its effect on the people it serves, and in every product launched on the confidence that problems can be fixed later, after the users have already become the experiment.

Ethics asks the question Victor avoids: who is responsible for what this creation becomes? That question becomes even more pressing when we move from the creature in the laboratory to the robot in the factory. Karel Capek introduced the word robot to the world through his play R.U.R., a story about artificial workers and the dehumanizing dreams of mechanized labor, and the word itself is rooted in the language of forced labor and servitude, which matters, because our earliest modern robot story was never only about machines. It was about work, control, exploitation, and the danger of treating sentient or humanlike beings as tools. Here is one of the stranger ironies of modern technology: we often imagine automation as liberation, yet the very language of the robot begins with labor under compulsion, and that irony should give us pause before we speak too easily of freedom. When we speak of AI doing the work no one wants to do, we ought to ask who gets freed and who gets displaced? When we speak of efficiency, we ought to ask efficient for whom, and when we speak of replacing human labor, we ought to ask whether we are removing drudgery or removing dignity, whether the human being is being served, or merely measured more aggressively than before.

The fear of artificial intelligence is not only the fear of being harmed by a machine. It is also the fear of becoming less human in a world organized entirely around machines, and that fear began long before today. The industrial age already taught people what it felt like to be swallowed by systems larger than themselves. Machines changed labor. Factories changed time. Clocks, shifts, wages, assembly lines, and management systems changed the rhythm of the body and the meaning of work itself, and a person who once saw the whole product of his labor might later perform one repeated task inside a system designed by someone else, for a purpose he did not choose and could not control. Industrial technology brought abundance, medicine, transportation, mass production, and comfort. It also brought exploitation, dangerous workplaces, pollution, and a new kind of human exhaustion, and the lesson here is not that industrial technology was evil but that power always arrives mixed with its cost, never given to us clean.

Might artificial intelligence do to judgment what industrial machines once did to labor? That sentence should slow us down, because if machines once reorganized the body, AI may reorganize the mind, and if factories once separated workers from the whole meaning of their work, AI may separate people from the whole meaning of their decisions. A person may no longer ask what do I think, but instead ask what did the system recommend. A manager may no longer ask what is fair, but instead ask what did the score produce, and a citizen may no longer ask who made this decision, but instead be told simply that the model indicated the outcome. This is not science fiction. It is the natural temptation of any system that makes complex decisions appear objective, and it is a temptation we are only now beginning to feel in full.

The history of AI begins with a bold dream of its own. In the 1950s, researchers at Dartmouth proposed a summer research project built on the idea that aspects of intelligence could be so precisely described that a machine could simulate them, and that proposal helped name artificial intelligence as a field. Alan Turing, before that, had already pushed the question of machine intelligence into public philosophy, asking whether machines could think and proposing a different way to examine the question through imitation and performance. From the very beginning, then, AI was never only engineering. It was philosophy wearing engineering clothes, and the questions beneath it, what is thought, what is intelligence, what does understanding actually require, what can imitation become, and what still separates a mind from a machine that merely behaves like one, did not begin with ChatGPT or modern generative AI. They were always inside the field, waiting patiently for the rest of us to catch up. What has changed is that ordinary people now feel these questions in the middle of an ordinary workday, when a student submits an essay assisted by AI, when a customer service answer sounds human but is not, when an image looks real but was generated, when a company wants to use AI to write, decide, screen, predict, rank, recommend, or replace.

That is why mythology and literature still matter here, giving us language for the emotional truth of the moment and reminding us that human beings have always known there is danger in creation without character. Prometheus reminds us that power changes the world. Icarus reminds us that limits matter. The golem reminds us that a helper with too narrow a mission can become dangerous, and Frankenstein reminds us that the maker remains responsible long after the making is done, while R.U.R. reminds us that a dream of artificial labor may carry within it the fear of dehumanization we have never fully resolved. None of these stories offers a complete AI policy, and that was never their purpose. Their purpose runs deeper still. They teach humility, and humility is not the enemy of innovation. Humility is what keeps innovation from becoming arrogance, and a culture that cannot be humbled by its own stories will eventually be humbled by its own consequences instead.

Some readers will say, at this point, that stories are not enough, and they are right. We need engineering standards, public policy, legal accountability, business practices, transparency, independent testing, and education. We need frameworks stronger than a mood and more durable than public anxiety. But stories still matter, because they tell us where the fear comes from, that people are not afraid simply because they fail to understand the technology. Sometimes people are afraid because they understand human nature far too well, having already seen what happens when power outruns wisdom, when leaders tell themselves the harm is temporary, the sacrifice necessary, the public incapable of understanding, the market certain to correct it, or history certain to forgive them. We have seen what happens when people build systems that reduce other people to categories, scores, outputs, risks, costs, or problems to be managed rather than persons to be known.

AI may be new, but the moral danger beneath it is not. The moral danger is ancient, the temptation to build before we become worthy builders, to create before we decide what love requires of the creator, to mistake intelligence for wisdom and power for permission, and that is why the fear deserves neither our mockery nor our surrender to it. It belongs to the human story. It is an alarm, and like every alarm, it can be annoying, excessive, or poorly timed, but it can also save a life if we let it. Will we listen well? If we listen poorly, fear becomes panic, and panic leads to bad policy, shallow slogans, and public confusion, making people reject useful tools because they fear imaginary ones, leaving them vulnerable to leaders who promise safety while quietly using the fear to consolidate control. If we listen well, fear becomes humility, humility becomes inquiry, inquiry becomes ethics, and ethics becomes design, governance, restraint, and responsibility in turn.

That is the path this book intends to follow, not fear as entertainment, but fear as instruction, fear as a doorway into wisdom rather than a wall that blocks it. To understand AI ethics, we first have to understand why the machine frightens us, and once we do, we discover something waiting underneath that fear, something we did not expect to find. We are not only afraid of the future. We are afraid of repeating the past with far more powerful tools than we have ever held before, and that fear, rightly understood, is not a reason to turn away. It is the very reason to look more closely.

Chapter 3: The Violence Was Here Before the Machine

There is a mistake we should avoid from the very beginning of this argument, and it is the mistake of pretending that artificial intelligence introduced danger into an otherwise innocent world. AI did not invent greed. It did not invent violence, racism, propaganda, surveillance, exploitation, corruption, war, or the ancient habit of treating certain people as less worthy than others, all of which human beings brought into the room long before any machine arrived to join us there. That does not excuse the machine, but it does clarify the problem, since a powerful tool entering a morally confused world will not automatically make that world wise. It may instead make the confusion faster, cheaper, more persuasive, and far more difficult to challenge, and the danger is not only that AI will become inhuman but that it will become an amplifier of the human failures we have never fully confronted in ourselves.

This is uncomfortable to sit with, because human beings prefer stories in which evil arrives from outside us, stories with dragons, monsters, demons, invaders, and villains, since these allow us to place danger somewhere beyond the ordinary human heart. But history keeps resisting that comfort, because the most destructive forces ever assembled were not built by machines acting alone. They were built by people who signed orders, designed systems, turned away from suffering, followed incentives, obeyed authority, and profited, people who told themselves they had no choice, that the harm was necessary, that the victims deserved it, that the future would justify the cost. Can the ethical conversation around AI remain honest if it is not first honest about humanity itself? It cannot, and any honest account must begin there.

We are capable of breathtaking beauty. We build hospitals, schools, libraries, churches, bridges, farms, homes, orchestras, and spacecraft, and we write poetry, nurse the sick, adopt children, forgive enemies, rebuild after storms, and cross oceans to help strangers we will never meet, planting trees we may never sit beneath. Any honest account of humanity must begin with the truth that people are capable of love, though it cannot end there, since we also build weapons capable of destroying entire cities. The United Nations describes nuclear weapons as uniquely dangerous because of their capacity for massive destruction, long term harm, and existential danger to humanity, and the Bulletin of the Atomic Scientists uses the Doomsday Clock as a public warning about how close human beings may be to catastrophe of our own making, warnings that existed long before generative AI became a household term. That should humble us, since long before the fear of a superintelligent machine, humanity had already created tools capable of ending human life at a scale previous generations could barely imagine, and the mushroom cloud remains one of the clearest symbols of the modern age precisely because it shows what happens when human intelligence separates itself from human restraint. It is science without enough mercy, precision without enough reverence, achievement standing in the shadow of its own annihilation.

Is this an argument against science? It is not. It is an argument against arrogance, since science can heal and science can destroy, technology can liberate and technology can enslave, data can reveal truth and data can be used to control, and efficiency can reduce suffering even as it removes the human pauses where compassion once entered. The moral direction of a tool depends, in every case, on the people, institutions, incentives, and limits placed around it, which is why AI ethics cannot be separated from human ethics. When people worry that AI might one day decide humanity is too violent to preserve, they are not inventing the violence. They are remembering it, remembering war, genocide, slavery, terrorism, abuse, corruption, ethnic hatred, mass displacement, and the endless human capacity to organize harm under the language of necessity.

Our World in Data tracks armed conflict and war through historical data, reminding us that the future of war and peace is not guaranteed, that conflict deaths can rise, that trends can reverse, and that peace has never been an automatic feature of history but has always depended on choices, institutions, norms, leadership, and restraint. AI enters that world exactly as it is, a world where people already know how to lie, where governments already know how to watch, where companies already know how to manipulate attention, where markets already reward scale before wisdom, where speed often defeats reflection, and where the vulnerable are so often tested on first and protected last of all. If we fail to understand that, we will misunderstand the danger entirely, since the machine is not the original sin. The machine is the accelerant, and an accelerant does not create the fire, though it can make the fire impossible to contain once it has already caught.

That is why we must be careful when we speak of AI becoming violent, since the more immediate question may be whether violent people, careless people, desperate people, greedy people, or powerful people will use AI to extend their own reach further than it could otherwise go. A person who wishes to deceive can create more convincing deception. A person who wishes to harass can multiply the harassment, and a person who wishes to avoid responsibility can simply say the system made the decision, a sentence that deserves our full attention, because one of the most dangerous ethical failures of this age will be the temptation to hide human choices behind machine outputs. The system recommended it. The algorithm flagged it. The model predicted it. These phrases can sound objective, but they may conceal a long chain of human decisions underneath their calm surface, since someone chose what data to collect, what to ignore, what to measure, what target to pursue, what risk to tolerate, what context to deploy within, whether to test for harm, whether affected people could appeal, whether there would be transparency, and whether profit, speed, public control, or human dignity would matter most in the end.

Artificial intelligence can make accountability more difficult, since the decision may feel diffused across teams, vendors, code, models, leaders, and institutions, but diffused responsibility is still responsibility, and no company, government, or individual should be permitted to place a system between themselves and the moral consequences of their own choices. This is one reason public frameworks matter. NIST's AI Risk Management Framework treats AI risk as something affecting individuals, organizations, and society together, which is the correct scale for this conversation, while

UNESCO's recommendation places human rights and human dignity at the center of AI ethics, the OECD AI Principles speak in the language of trustworthy AI, human rights, and democratic values, and the European Union's AI Act shows that governments are already attempting to classify and govern AI systems according to risk. These sources do not solve the whole problem, but they show that the world has begun to recognize a basic truth, that AI is too powerful to be left only to enthusiasm, marketing, or private judgment.

Will frameworks alone save us? They will not, since a principle written on paper is not the same as a principle lived under pressure, and every organization says it values people, every government says it serves the public, every company says it cares about trust, every leader knows the language of responsibility when the cameras are on. The real question is what happens when values become expensive, when the ethical choice slows growth, reduces profit, limits power, or requires admitting a mistake in public, and that is where character appears. AI will test character, because it will offer extraordinary temptation, tempting businesses to replace service with simulation, governments to replace judgment with automated control, schools to replace learning with answer production, individuals to replace reflection with convenience, and leaders to say yes before they understand what yes will cost. The danger is not only misuse by bad actors. The danger is ordinary use by people who never pause long enough to ask whether ordinary use is forming them into better or worse people over time.

A society can be degraded by nothing more than convenience, not all at once, not in a dramatic collapse, but slowly, almost politely, as we let the tool finish the sentence, then let it write the message, then let it decide what matters, then let it remember for us, then let it judge for us, until we forget what it once felt like to struggle with a thought until it finally became our own. That is not doom in the cinematic sense, but it is still a kind of loss, and a real one.

Part of this book will later argue that AI can help human beings flourish, and I believe that fully. I do not believe the ethical answer is to reject the tool. I use AI. I teach about AI, and I believe business owners, leaders, educators, and ordinary people can use these tools to become more effective, more creative, more organized, and more capable of serving others, but belief in the promise should never require blindness to the danger. A tool can help us become better only if we use it toward what is actually better, which raises the hard question underneath everything else in this chapter: what, then, is better?

If better means only faster, AI will make us faster. If better means only cheaper, AI will make things cheaper. If better means only more profitable, AI will help profit seek every opening it can find, and if better means more control, AI will become a control system. But if better means more human, more truthful, more merciful, more just, more creative, more responsible, and more aligned with the dignity of people, then AI must be built and governed around those aims from the very beginning, not added to them later as an afterthought.

The Universal Declaration of Human Rights begins with the idea that human beings are born free and equal in dignity and rights, a moral starting point that matters because AI systems will increasingly touch the conditions under which people live, work, learn, speak, receive care, and participate in society. Human dignity cannot be an optional setting. It cannot be a public relations phrase, and it cannot be something we remember only after the system has already caused harm. Dignity must become a design requirement, a leadership requirement, a policy requirement, and a personal requirement all at once, woven into the foundation rather than painted on afterward.

This is where the fear of doom becomes genuinely useful, forcing us to ask whether we are designing systems that preserve humanity, or merely systems that perform impressive tasks, whether we are creating intelligence in service of wisdom, or intelligence in service of whatever pays best, whether the people building the future are also being formed into the kind of people who can be trusted with it. Will the fear of AI destroying humanity ever become literal? I do not know, and no honest person knows with perfect certainty, but the fear that human beings will use powerful tools in harmful ways is not speculative. It is history, plainly written and repeated across every century we have kept record of.

That is why this chapter matters. Before we ask whether AI will become violent, we must ask what violence we are already teaching it through our data, our systems, our incentives, and our example. Before we ask whether AI will become greedy, we must ask what forms of greed we are rewarding, and before we ask whether AI will become biased, we must ask what biases already live inside the records we have kept and the institutions we have trusted for generations without question. Before we ask whether AI will value human beings, we must ask whether we have built a society that values human beings when they are poor, inconvenient, old, disabled, unprofitable, different, angry, slow, confused, or unable to contribute in ways the market can easily measure.

AI is a mirror, though not a passive one, since a mirror only reflects, while AI can reflect, repeat, amplify, automate, and persuade, taking the patterns of our world and returning them to us with speed and confidence we may not have earned. That is what makes it powerful, and that is precisely what makes it dangerous, for if we train systems on the world as it is, without a clear vision of the world as it should be, we should not be surprised when they reproduce the very failures we claim to oppose in public while practicing quietly in private.

The point of this chapter is not that humanity is hopeless. The point is that humanity is unfinished, still learning how to live with power, how to see each other fully, how to govern ourselves, how to tell the truth, how to restrain appetite, how to use wealth with justice, how to protect the weak, how to forgive, and how to build systems that do not crush the very people they claim to serve. AI arrives in the middle of that unfinished work, which is precisely why ethics cannot wait until later.

Later is what people say when they want the benefit now and the responsibility postponed indefinitely. Later is what organizations say when the launch date matters more than the harm assessment, and later is what governments say when power is available before accountability has been built. Later is what a society says when it wants the comfort of progress without the burden of wisdom, and later has a cost, since the systems we build now will shape the habits, expectations, markets, laws, and cultural norms of the future. Once a technology becomes embedded, it grows harder to question. Once a business model forms around extraction, it grows harder to reform, and once a government becomes accustomed to automated authority, it grows harder to restore human discretion. Once citizens become dependent on convenience, it grows harder still to recover the judgment they once carried freely.

This is why the fear must be taken seriously, not because fear should rule us, but because fear is telling us the stakes are real. We do not need to believe every apocalyptic prediction to understand that AI demands a moral response. We only need to look honestly at what human beings have already done with power, at what companies do under pressure, what governments do when afraid, what individuals do when anonymous, what markets do when unrestrained, and what institutions do when no one can see inside them at all.

The violence was here before the machine. The greed was here before the machine. And the dream was here too, a sentence that matters, since we should not end Part One in despair. Human history is not only a record of harm. It is also a record of conscience, and we have ended cruelties that once seemed permanent, expanded rights, built institutions to restrain power, and written declarations that name the dignity of every person, creating tools that heal, feed, teach, connect, and protect while watching human beings risk their own lives for strangers they had never met.

Does the fact that we are dangerous mean we are doomed? It does not. It means we must become more honest, and ethics in AI begins with precisely that honesty, with the admission that the machine will not arrive in a moral vacuum but will arrive inside the world we have already made, carrying forward, unless we are careful, the very things we most need to overcome. The question is not only whether we can control AI. The question is whether we can control ourselves well enough to build AI that deserves to be trusted, and whether the page in front of us can still be written differently than the one we fear.

Closing Reflection: The Fear Is Not the Enemy

Fear is not the enemy of this book. Fear is where the conversation begins, because fear tells us something important is at stake, and we should not mock people for being afraid of artificial intelligence, nor dismiss them as uninformed, old fashioned, dramatic, or unwilling to change. Some fears are exaggerated. Some are poorly aimed. Some are built more from movies than from facts, but beneath the surface, many of these fears are pointing toward real moral questions worth asking directly rather than waving away.

The anxiety grows whenever machines can imitate us, whenever tools begin making decisions on our behalf, whenever power moves faster than accountability, whenever the people most affected by a system cannot understand it or challenge it, and whenever humanity builds something powerful before agreeing on what should guide it. Part One began with doom, because doom gets our attention, and it looked backward, because our fear of AI is part of a far older human story, ending by naming an uncomfortable truth: artificial intelligence did not introduce violence, greed, control, or dehumanization into the world. Those things were already here, waiting quietly for a faster way to travel.

Does that leave us hopeless? It does not. It leaves us responsible, and if the machine is a mirror, the answer is not to smash the mirror. The answer is to become worthy of what we are building, page by page, choice by choice, while the ending still remains unwritten and the pen, remarkably, is still in our hand.

The next part of this book turns toward the dream, and not a shallow one, not a marketing dream, and not a fantasy in which technology magically solves the human heart, but a serious dream, a future where AI remains within human control and helps us cure disease, reduce unnecessary suffering, improve education, protect the planet, remove the weight of meaningless labor, and give people more room to become what they were created to be. The promise is real. But will the promise protect itself. It will not, and that is why we had to start with fear, and why we can now, finally, begin to hope.

PART TWO

The Dream

Hope deserves to speak, but hope has to tell the truth.

Chapter 4: The Utopian Promise

The dream is not that the machine becomes our master. The dream is that the machine helps us become more human, and after fear has had the first word, hope deserves to speak in turn.

There is a version of the future where artificial intelligence does not destroy us, divide us, or quietly turn us into dependents of systems we no longer understand, a version where one of the most powerful technologies in human history remains within our control and helps us become more like the people we say we want to be. That future is not silly, not childish, and not the blind optimism of people who have never studied power, greed, failure, or human nature. It is a serious possibility, deserving to be taken seriously, because the same tool that can be used to manipulate a population can also be used to detect disease, the same kind of system that can recommend a product can also help a child with a disability communicate with the world, and the same kind of machine intelligence that can generate noise can also search through scientific possibilities that would take human researchers entire lifetimes to test on their own.

The dream is not that artificial intelligence becomes good on its own. The dream is that human beings become wise enough to build, guide, limit, and use it for good, and we must be honest about the scale of this moment, since artificial intelligence is not another app on a phone, not simply another software upgrade, and not merely a faster version of search. It is a general purpose force that can touch medicine, education, government, finance, transportation, military planning, agriculture, entertainment, law, energy, parenting, business operations, and the ordinary decisions of daily life, and that does not make it magic. It does make it historically important, and when a tool begins to touch everything, the moral question grows larger than the tool itself.

The printing press changed who could read and who could challenge power. Electricity changed the rhythm of the day. The internet changed how information travels, and artificial intelligence may change what it means to think with a tool, to make decisions with a tool, to govern with a tool, and to imagine the future with a tool in hand rather than only a mind. That is why the utopian vision matters, since if we speak only of doom, we make fear the architect of the future. If we speak only of profit, we make money the architect, and if we speak only of efficiency, we make speed the architect, but if we speak honestly about human flourishing, we at least give ourselves a chance to ask what this technology should be for in the first place, before we simply hand it the keys.

The best version of AI begins with a simple idea: human beings should matter more because of it, not less, and in medicine, the promise is easy enough to see. A doctor can only hold so much information in one mind. A nurse can only monitor so many signals at once, and a researcher can only read so many papers, compare so many molecules, and test so many possibilities before a lifetime runs out. AI does not need to replace the doctor to change medicine. It can help notice patterns, assist in diagnosis, help researchers move faster, and help public health officials see danger earlier, supporting discovery in drug development and scientific research when paired with careful governance and human accountability. The World Health Organization has already treated AI in health as both an opportunity and a serious ethical responsibility, especially where public benefit, equity, safety, and oversight are concerned.

The example of protein structure prediction gives us a glimpse of what this can look like in practice. Proteins are part of the machinery of life, and understanding their shapes matters for biology, disease research, and medicine, and AlphaFold showed that machine learning could help predict protein structures at a scale and speed that changed scientific access to that problem entirely, opening predictions for millions of protein structures to researchers around the world. This does not mean AI cures disease by itself, or understands suffering, or should make medical decisions alone, but it does mean that when human knowledge, human care, and machine capacity are joined carefully, new possibilities do appear where none existed before. A parent sitting beside a hospital bed does not much care whether the breakthrough came from a person alone or from a person using a machine. That parent cares only whether the child gets to live, and that is where the dream becomes real, not in speeches, and not in marketing, but in a family that gets more time, a patient who gets an answer sooner, and a disease that finally becomes treatable because we saw what we could not see before.

The same is true in education, where the dream is not that AI replaces teachers, since a teacher is not merely a content delivery system. A good teacher notices frustration, discouragement, curiosity, embarrassment, effort, and growth, and sees when a student is trying to disappear into the back of the room, calling something out of a young person that the young person does not yet know is there. AI cannot love a student, but it may help a teacher reach more students, translate material, adjust practice to a student who learns differently, and help a child with limited access to tutoring receive support that would otherwise be unavailable to them. That is the right kind of promise, not replacement but support, not the removal of human relationship but its strengthening.

This same promise appears in disability access, where technology has always carried a moral dimension, since design decides who gets included and who is quietly left outside. A curb cut helps the person using a wheelchair, but it also helps the parent pushing a stroller, the worker moving a cart, and the child learning to ride a bike, and the same principle applies to AI. Speech tools, vision tools, captioning, navigation support, adaptive interfaces, and learning support can give people access to work, creativity, communication, and independence that may have been far harder to reach before these tools existed. Microsoft has framed AI and accessibility around daily tasks, work, creativity, and innovation rooted in disability communities themselves. Inclusion is not charity. Inclusion is design finally telling the truth, since human ability was never one fixed thing to begin with. We all live with limits, some visible and some private, some arriving at birth and some through injury, aging, illness, grief, or exhaustion, and a humane technology does not treat the average user as the only user. It widens the door, and if AI helps widen that door, it is serving humanity in the truest sense of the word.

Agriculture offers another picture of the dream. Food is one of the oldest human problems, since we must grow it, transport it, protect it, price it, and share it, and farmers already live with weather, soil, pests, markets, labor pressure, equipment cost, and unpredictable risk that no algorithm can fully remove. AI and digital agriculture may help with precision farming, climate smart agriculture, supply chain optimization, market access, and food security, and the Food and Agriculture Organization has described digital technologies and artificial intelligence as opportunities to transform agrifood systems with greater efficiency, sustainability, and resilience. This should not become a fantasy where machines do everything and farmers become irrelevant. The point is not to erase the farmer but to give the farmer better tools, since a good tool does not make the person smaller. It extends the reach of the person holding it.

The same could be true for climate, energy, logistics, and the stewardship of the planet itself, since AI can help model complex systems, reduce waste, manage traffic, predict demand, analyze climate patterns, improve grid efficiency, and find better ways to use limited resources than we have managed on our own. International organizations have already connected AI with global challenges and sustainable development goals, while warning that those benefits depend on governance, access, and responsibility rather than arriving automatically.

This is the part of the utopian vision we should not be embarrassed to say aloud. Maybe AI can help us waste less, heal more, and learn faster. Maybe it can help us make better use of the resources we already have, and help people who have been left out finally gain access to tools, knowledge, care, and opportunity they were long denied. Maybe, above all, it can help make the world a little less cruel, and that last one is the one that matters most, not perfect, not controlled entirely by machines, not scrubbed clean of disagreement, hardship, ambition, competition, or sorrow, simply less cruel. A world where fewer people are crushed by work that gives them nothing back. A world where fewer people are ignored by systems too busy to notice them, where fewer people are denied care because no one had time to read the file, where fewer children fall behind because no one had enough resources to help them, and where fewer people with disabilities are told the world was never built for them in the first place.

Is that worth dreaming about? It is, though the dream becomes dangerous the moment it forgets its own condition, since AI can help us build a better world only if human beings decide that a better world is actually the goal, and that sentence sounds simple until we sit with it a while. A company may say better and mean more profitable. A government may say better and mean more controllable. A consumer may say better and mean more convenient, and a military may say better and mean more dominant, while a platform may say better and mean more addictive, and a politician may say better and mean more persuasive. A family may say better and mean more time together, a teacher may say better and mean deeper learning, and a doctor may say better and mean fewer preventable deaths. The same word, it turns out, can hide a thousand different motives beneath its surface.

That is why the dream cannot be left to slogans. We need a moral definition, since a better world cannot simply mean a faster world, a richer world, or a world where fewer people are needed. It cannot simply mean a world where every process is optimized and every inconvenience removed, but must mean instead that human beings are treated with more dignity, more fairness, more care, more freedom, more opportunity, and more responsibility than before. This is where ethics begins to move from theory into design, since if AI is going to help us become better, we have to define better before the system begins helping us pursue whatever we happen to reward, because AI will not only follow our words. It will follow our incentives. If we reward attention at any cost, it will help capture attention. If we reward profit at any cost, it will help extract profit, and if we reward power at any cost, it will help concentrate power, but if we reward truth, care, transparency, dignity, and human flourishing instead, we will have given the system a genuinely different direction to travel toward.

The dream requires control, though not control in the sense of domination but control in the sense of stewardship, the kind of adult responsibility that never permits a technology this powerful to drift wherever the market, the state, or the loudest voice happens to take it. Human centered AI is not anti technology. It is the only way the technology earns trust in the first place. Some of the builders have argued for designs that combine strong automation with strong human control, allowing technology to improve human performance while preserving safety, reliability, and responsibility. That is the shape of the promise: powerful tools, joined to human control, joined to human dignity, opening a wider door toward a better life than the one we currently live.

The dream is not that AI rescues us from being human. The dream is that AI helps remove some of the burdens that keep us from living humanly, and there is a real difference between the two, since to be human is not merely to survive. It is to love, raise children, build homes, plant gardens, mourn the dead, forgive the living, create beauty, serve neighbors, tell stories, and search for meaning that outlasts us. If AI gives people more space for those things, it can be part of a more humane future, but if it gives people more convenience while quietly stealing judgment, privacy, relationship, responsibility, and purpose, the dream becomes only a softer version of the nightmare we began with in the first chapter of this book.

That is why the utopian picture must remain both convincing and conditional. Yes, AI can help us become better, but only if we build it around the best of us, not the easiest part, not the greediest part, not the angriest part, and not the part that wants to win at the expense of everyone else in the room. The best of us is the part that sees a stranger and still recognizes a person, that knows a child is not a data point, that understands the elderly are not inefficiencies, that believes the poor are not failed inputs, and that knows the planet is not merely inventory to be managed. It is the part that understands power is only ever honorable when it is used with restraint, and that is the version of the dream worth fighting for, a dream we are still, even now, allowed to write.

Chapter 5: The Question Beneath the Dream

The hardest part may not be building the machine. The hardest part may be agreeing on what we are asking it to help us become, since the dream forces a question older than artificial intelligence itself: what is good? Not what is useful, not what is profitable, not what is popular, and not what is efficient, but good, a word small enough to say quickly and large enough to divide civilizations that have never agreed on much else.

That is the question beneath the dream, sitting underneath every speech about innovation, every policy paper about safety, every corporate promise about responsible AI, every government hearing, every classroom debate, every product launch, and every family conversation about whether children should use these tools at all. What, in the end, are we actually trying to build, and a better world sounds easy until people begin defining better in their own particular terms? For one person, better means safer. For another, freer, or wealthier, or more equal, or more obedient, or more efficient, or more creative, or more traditional, or more connected, or simply being left alone at last. These differences are not small. They shape laws, markets, schools, families, and nations, and they will shape AI in exactly the same way, whether we notice it happening or not.

This is why the ethics conversation cannot begin with the machine. It has to begin with us, since artificial intelligence forces humanity to confront a problem we have always carried quietly within ourselves. We do not fully agree on what human beings are for. We do not fully agree on what society owes the individual, or what the individual owes society in return, and we do not fully agree on where freedom ends and responsibility begins, on how much inequality is tolerable, on what privacy means in a world where every movement can be tracked, or on how much control a government should hold, how much influence a corporation should wield, or how much convenience a person should trade for ease. Did AI create those disagreements? It did not. It exposes them, makes them faster, and makes them far harder to ignore than they were before.

This is why we must be careful when people say AI should be aligned with human values, since it sounds right, and may even be necessary, but whose values, which humans, which culture, which government, which company, which religion, which ideology, which class, which generation, which language, which history. The phrase human values sounds unified from a distance, but up close it grows complicated quickly, since some values are deeply shared across cultures, human dignity, the protection of children, the desire to be treated fairly, the need for safety, the importance of truth, the pain of humiliation, the longing to belong, and the desire to raise a family in peace, values that are not identical in every place but are recognizable nearly everywhere we look. Other values remain genuinely contested, the proper role of religion, the limits of speech, the meaning of equality, the balance between security and privacy, the rights of the individual against the demands of the group, and the relationship between tradition and progress, questions that do not disappear because a machine enters the room but grow only more urgent because of it.

If an AI system is built in one culture and deployed in another, whose assumptions travel with it. If a company defines safety in a way that protects its business but restricts the user, is that safety, or is it control dressed in gentler language. If a government defines misinformation in a way that protects public order but silences dissent, is that truth, or is it power wearing a more presentable face, and if a platform defines engagement as success, does it matter whether that engagement comes from wisdom, or from outrage, addiction, and fear. These are not technical questions wearing moral clothing. They are moral questions that technology has made impossible to avoid any longer, however much we might prefer to.

International frameworks have tried to build common ground here. UNESCO placed human dignity, human rights, safety, fairness, proportionality, and social responsibility near the center of AI ethics, while the OECD framed its AI principles around trustworthy systems that respect human rights and democratic values, and NIST built a risk management framework asking organizations to consider AI risk across people, organizations, and society together rather than in isolation. These frameworks matter because they show the world is not beginning from nothing. We have language, we have principles, and we have institutions trying to build a shared vocabulary out of genuine disagreement.

But are principles enough if they remain posters on a wall. They are not, since a principle that does not change a budget is not yet a principle, and a principle that does not change product design is not yet a principle either. A principle that cannot survive pressure from investors, voters, competitors, or military urgency is not yet a principle at all. It is branding, and the real test of AI ethics will never arrive when the choice is easy. It will arrive when the ethical choice costs something real, and the dream becomes serious only when people are willing to make those costly choices. A company may need to slow a product because it could harm vulnerable users. A government may need to refuse a surveillance tool because it makes authority too easy to abuse, and a school may need to protect student thinking even when AI can produce cleaner assignments faster than any student could write alone. A business may need to disclose AI use even when silence would be easier, and a developer may need to say no to a profitable system that should never exist in the first place. All of us may need to refuse the temptation to turn every human weakness into a revenue model dressed up as innovation.

That is where ethics becomes real, and the dream requires agreement, though not perfect agreement, since perfect agreement is neither possible nor, perhaps, even desirable, and a world where everyone is forced into one moral vocabulary can become its own kind of tyranny. What we need instead is a moral floor, since a floor is not the whole house? It is what keeps the house from collapsing, and the moral floor for AI must begin with the truth that every human being has dignity, not usefulness, not market value, not data value, and not productivity, but dignity itself. The Universal Declaration of Human Rights begins from that basic moral posture, and modern AI ethics cannot afford to move backward from it, whatever pressures push in that direction.

From that floor, certain commitments follow naturally, in the way roots follow from soil. A person should not be reduced to a profile. A child should not be treated as an experimental market, and a worker should not be managed by invisible judgment with no path of appeal available to them. A patient should not be denied care because a system reached a conclusion no human being can explain, and a citizen should not lose rights because authority hid behind automation rather than standing before the public directly. A community should not bear the harm of a system that was never tested on people like them at all, and these are not abstract concerns. They are the shape of dignity under the particular pressure of new technology.

The second part of the moral floor is human agency. People must remain participants in their own lives, and AI should not slowly train us to stop choosing, stop thinking, stop questioning, or stop accepting responsibility for what we decide. There is a dangerous kind of comfort in being relieved of decision making entirely, since the system knows what you want, what you should buy, what news you should see, which applicant is best, which tenant is risky, which student is struggling, and which patient is worth the cost, and the more often we hear that, the more important it becomes to ask whether a person can still question the answer at all, or whether the answer has quietly become final.

Trustworthy AI cannot only be accurate. It has to be contestable, leaving room for correction, appeal, context, mercy, and human judgment to operate freely within it. NIST describes trustworthy AI as involving multiple characteristics of value to interested parties, which points toward a larger truth: trust is not created by performance alone. It is created by accountability, transparency, safety, fairness, privacy, resilience, and the ability of people to understand and challenge what affects them directly.

The third part of the moral floor is responsibility. No one should be able to blame the machine and walk away, not the developer, not the executive, not the government official, not the school, not the hospital, not the platform, and not the business owner standing behind all of them. AI should not become the new language of cowardice, the system made the decision, the model flagged the risk, the algorithm recommended the denial, sentences that do not settle responsibility no matter how confidently they are spoken. A system does not sign a contract with the public. A system does not take an oath, and a system does not stand in front of a parent, a patient, a worker, a tenant, a citizen, or a child to accept moral accountability for what has happened to them. People do. Institutions do. Leaders do, and that distinction has to remain clear no matter how complex the technology becomes.

The fourth part of the moral floor is restraint. A civilization that can build more than it can morally govern is a civilization in danger, and this is not new. We learned it with weapons, with industry, with chemicals, with data collection, with financial engineering, and with social media long before AI arrived on the scene. Ability is not permission. A thing can be possible and still be wrong, and this may be one of the hardest truths for an innovative culture to accept, since we like progress, we like building, we like speed, and we like being first, rewarding whoever arrives before everyone else regardless of what they leave behind. But a tool that affects the human mind, the public square, the economy, education, medicine, government authority, and personal identity cannot be governed only by the energy of invention. It needs the discipline of restraint standing beside it, and that does not mean fear should stop progress, since fear alone is a terrible builder. But wisdom does sometimes say wait, sometimes says test again, show your work, involve the people who will carry the consequences, and sometimes wisdom simply says no. A moral society needs that word, not because we hate technology, but because we love people more than we love the technology itself.

The fifth part of the moral floor is participation. The people affected by AI should have a voice in the systems that affect them, especially children, people with disabilities, workers, poor communities, minority communities, patients, students, and citizens navigating government services that were never designed with them in mind. Systems designed only by the powerful often carry the particular blindness of the powerful, invisible to those who built them but obvious to everyone standing beneath their weight. UNICEF has warned that AI systems affecting children need policies and designs that uphold children's rights, especially as AI becomes more present in their lives with each passing year. That principle extends further still, since when AI touches a vulnerable group, that group should never be treated as an afterthought. They should be part of the conversation from its very start, not consulted only after the design is already finished.

The dream of AI cannot be handed down from a boardroom, a laboratory, a capital city, or a founder's office as though the rest of humanity were only a user base waiting to be managed. The dream has to be shared, which means the question beneath the dream is not only philosophical. It is civic, and it asks simply, how do we decide together. We start by admitting we cannot outsource the definition of good to the machine, since AI can help analyze options, model risks, and reveal patterns we might otherwise miss, but it cannot carry the moral burden for us no matter how sophisticated it becomes. We also start by refusing to let one institution define the future alone, not government alone, not business alone, not academia alone, not technologists alone, not activists alone, and not investors alone, since a future that affects everyone needs more than expert control. It needs expert knowledge joined with public accountability, technical wisdom joined with moral wisdom, speed balanced by deliberation, and people close to the problem joined with people trained to see the whole system at once.

Is that slower than simply building and launching? It is, and that is good, since some things should be slower, and a society that cannot slow down for wisdom will eventually be slowed down by consequences instead, consequences that rarely arrive on a schedule of our choosing. None of this means every small AI use requires a global council. We do not need a hearing every time someone uses a writing tool, summarizes a meeting, organizes a schedule, or drafts a grocery list, but the more a system affects rights, freedom, money, health, housing, employment, education, safety, public services, or the formation of children, the stronger the ethical requirements must become. The deeper the impact, the higher the burden, and that rule is simple enough for ordinary people to understand and strong enough to guide real policy across every setting where it might apply.

A spelling suggestion is not the same as a sentencing recommendation. A playlist is not the same as a hiring decision, and a customer service chatbot is not the same as a medical triage tool, since different uses carry different moral weight, and that is why the dream has to become disciplined rather than merely hopeful in the abstract. We cannot agree on everything, but can we agree that some areas require a higher standard. We can, that people should know when AI affects important decisions, that humans must remain accountable, that children deserve special protection, that privacy matters, that human dignity is not a feature to be added later, that systems should be tested before they are trusted, and that a person harmed by automation should have a real path to challenge it. If we cannot agree even on that much, then the problem is not AI. The problem is us, and the machine is not the only thing that needs alignment. Humanity does too, perhaps more urgently.

That is the real challenge, and it is also, if we are honest, the real hope resting quietly beneath it. The best future will not arrive because AI becomes powerful. It will arrive only if human beings become clear about what matters, about what should never be traded, about who must be protected, about the limits of profit and the limits of authority, and about the value of a person even when that person is not efficient by any measure we know how to keep. The dream is possible, but it needs agreement deep enough to guide power, not perfect agreement, but moral agreement, the kind that says every person matters, that the vulnerable are not acceptable collateral, that truth is not optional, and that no system becomes too large to question. The kind that understands the future is not merely something we predict. It is something we are still, together, responsible for writing with whatever time remains to us.

Chapter 6: What Makes Us Human?

If we do not know what a human being is worth, we will eventually build systems that treat people as though they are worth only what they produce, and before we can decide what AI should do, we have to remember what human beings are in the first place. That may sound strange, since we live as humans every day, waking in human bodies, carrying human histories, loving with human hearts and worrying with human minds, raising children, losing people, making mistakes, carrying regrets, telling stories, building homes, starting over, and trying again more times than we can count. We know what being human feels like, because we cannot escape it even for a moment, but knowing what it feels like is not the same as protecting what it means, and that distinction matters more than it first appears.

Modern life has trained us to measure people constantly, output, attention, grades, wealth, productivity, credit, engagement, health markers, performance, speed, and influence, counting how many people followed, clicked, bought, watched, liked, opened, shared, or stayed, and measurement can be useful, but it can also become a sickness when it forgets what it was meant to serve. A human being is not the sum of what can be measured, and this matters because AI is built on measurement. It needs data. It looks for patterns. It ranks, predicts, classifies, generates, recommends, and optimizes, and all of that can be powerful and genuinely helpful, but it can also become dangerous the moment we begin to believe that the measurable parts of a person are the whole person, which they never are and never will be.

A child is more than test scores. A worker is more than productivity, and a patient is more than a chart, just as a tenant is more than a risk profile and a grieving person is more than a behavior to be corrected. A human life is not a spreadsheet with skin stretched over it, and if AI ethics begins anywhere, it must begin there, with human beings holding dignity before they produce anything at all. The Universal Declaration of Human Rights gives language to this idea at a global level, and modern AI ethics should not treat that dignity as optional or negotiable under pressure. The danger of a highly optimized world is that it may become very good at forgetting this simple truth, since a system can be accurate and still be cruel, efficient and still be unjust, profitable and still be corrosive, reducing cost while increasing humiliation, processing more people while seeing fewer of them as individuals at all.

This is why we need a deeper account of humanity than mere usefulness can provide. Human beings are creatures of meaning. We do not merely want to exist. We want our existence to matter, our pain to mean something, our work to serve something larger than ourselves, and our children to carry something forward long after we are gone. We want our love to be received, and we want our lives to be more than a sequence of needs being met one after another until they stop. This is one reason the utopian version of AI cannot simply be a world where machines give us everything we want, since a life of endless consumption is not the same as a meaningful life, and a life with no hunger, no work, no pressure, and no danger might sound peaceful at first, though human beings were never built only for comfort. We are also built for purpose, needing something to care about, something to build, someone to serve, and difficulties that develop us rather than destroy us in the process.

This is where the dream of AI must be handled with real care, since removing meaningless labor is good, but removing every form of effort is not, and reducing preventable suffering is good, while eliminating every challenge is not. Giving people more time is good, though time without meaning can quietly become emptiness in its place, and the future should not be designed to make human beings passive. It should be designed to make human beings more capable than before. Aristotle's idea of human flourishing helps us here, since a good life is not merely pleasure or comfort but involves virtue, choice, excellence, habit, and the slow development of character over a lifetime. That ancient insight matters in a modern AI conversation, because it reminds us that human life is not only about outcomes. It is about formation, about who we are becoming rather than merely what we are getting, and that is a very different question from the one most systems are built to answer.

AI can help us get things faster, answer questions faster, generate images faster, write messages faster, analyze data faster, build plans faster, and solve problems faster than we ever could alone, but is faster always deeper? It is not, since a person can get an answer without gaining wisdom, generate a paragraph without learning to think, simulate friendship without practicing it, automate an apology without becoming repentant, produce art without being changed by the making of it, and receive advice without ever developing judgment of their own. None of this means we should reject the tool. It means we should understand plainly what the tool cannot do for us, since AI can assist the work of becoming human. It cannot do the becoming for us, and no amount of speed will ever substitute for that slower, harder work.

This is especially important in education, where children need help, access, adaptation, and support that AI can genuinely provide, but children also need struggle, need to wrestle with a sentence until the thought becomes their own, to sit with confusion long enough to develop patience, to make mistakes and learn how to recover from them, and to feel the pride that comes only from doing something difficult honestly and without shortcut. If AI removes every hard edge from learning, it risks removing some of the formation that learning provides in the first place, and the same holds true for adults. A business owner who uses AI to think more clearly may become better at what she does. A business owner who uses AI to avoid thinking may become weaker instead, and a leader who uses AI to prepare more carefully may serve better, while a leader who uses AI to sound wise without becoming wise may become more dangerous than before, precisely because the appearance of wisdom can travel further than wisdom itself.

Tools shape the people who use them. A hammer focuses our momentum. A calculator changes mental arithmetic, and a map changes how we navigate, while an iPhone changes memory, attention, and social behavior in ways we are only beginning to understand. AI will shape judgment, creativity, patience, confidence, and dependence in exactly the same way, which means the ethical question here is deeply personal, not only public. We cannot only ask what AI will do to society. We must ask what it will do to us, individually, in the quiet hours when no one else is watching?

The same concern applies to love and relationships. AI can simulate attention, respond quickly, remember preferences, use caring language, offer affirmation, and listen without ever growing tired of listening, and for a lonely person, that might feel powerful, even provide some genuine comfort in a difficult season. But is human love only a response? It is not, since love costs something, has a body, gets interrupted, forgives, waits, sacrifices, changes plans, and stands beside a hospital bed when standing there is the hardest thing in the world to do. Love shows up precisely when it would be easier not to, and it is not merely the experience of being understood. It is the responsibility of belonging to another person, fully and without an exit clause.

AI may imitate the language of care, but it does not carry the burden of care the way a person does, and that distinction will matter more and more as artificial companionship becomes more convincing with each new model. The ethical danger is not only that people will be fooled. It is that people may begin to prefer relationships that ask nothing from them, since a companion that never needs patience, forgiveness, compromise, or sacrifice may feel easier than a human relationship ever could. Is easier always better? No, since the best parts of us are so often formed by people who are not easy at all. Children are not easy. Marriage is not easy. Friendship is not always easy, and neither are aging parents, community, or forgiveness, but a life without those burdens would also become a life without the depth they create in us over time.

That is why AI should never be allowed to redefine human connection as a problem to be solved. Loneliness is a real wound, and AI may help people through certain moments of it, but the final goal must remain human connection, not a permanent artificial replacement standing in its place.

We also need to speak honestly about conscience, since human beings do not only calculate. We wrestle, knowing what it feels like to be pulled between self interest and duty, to say the wrong thing and wish we could take it back, to feel guilt, to feel shame, to know what courage costs and what forgiveness requires of the one who offers it. A machine can analyze ethical frameworks, generate arguments, identify inconsistencies, and predict consequences with impressive speed, but it does not have a conscience in the human sense. It does not bear guilt. It does not repent. It does not grow in virtue through suffering. It doesn't know failure, correction, and grace the way a person does across an entire lifetime.

That does not make AI useless in ethical reasoning. It can help us see blind spots, compare options, and surface consequences we might otherwise miss entirely, but it cannot become the moral center of a life. The moral center has to remain human, and this is where human oversight becomes far more than a policy phrase. It becomes a statement about the very nature of responsibility, and international AI frameworks often emphasize human rights, human oversight, safety, and accountability, because decisions affecting people cannot be morally abandoned to a system, however capable that system appears to be. Someone has to answer. It must be able to correct, apologize, make restitution, and learn from the mistake, and that someone has to be a person capable of all four.

AI may become intelligent in ways that genuinely amaze us, better than us at many tasks, capable of processing more information and detecting more patterns than any single human being ever could manage alone, but is intelligence the same as wisdom. It is not, since wisdom involves judgment, judgment involves values, and values involve love, fear, memory, culture, consequence, and responsibility, while responsibility involves a person who can actually be held to account for what they have chosen. That is why a humane AI future must protect the slow, difficult, beautiful parts of human life: conversation, family, craft, prayer, rest, study, service, grief, friendship, citizenship, and wonder. Protect the embodied life itself, the feel of dirt in a garden, the sound of a child laughing in the next room, the smell of food cooking for people we love, the ache in our bones after honest work, the unfillable silence after loss, the courage to tell the truth when lying would protect us, and the discipline to keep a promise after the emotion that made it has long since passed into memory.

Are these inefficiencies? Not at all. They are the substance of human life, and a world built only for optimization might fail to recognize them at all, mistaking the very things that make us human for waste to be trimmed away. The danger, then, is not that AI becomes a person or even . The danger is that people begin accepting systems that cannot understand personhood as the final authority over persons, and we must build against that danger deliberately, testing every major system against a single basic question: does this help human beings live with more dignity, agency, responsibility, truth, relationship, and meaning? If yes, we should explore it carefully. If no, we should ask ourselves why we are building it at all, and stop pretending the answer does not matter simply because the technology is impressive.

That question will not answer every dispute or remove every risk, but it gives us a compass, and the future does not need AI that merely makes us faster. It needs AI that helps us become wiser, remembering always that human beings were never meant to be reduced to outputs, profiles, scores, predictions, or transactions. We are not machines with appetites. We are not data waiting to be extracted, nor problems waiting to be optimized. We are souls, bodies, stories, families, neighbors, citizens, and stewards of a planet we did not create and do not own, and if AI helps us remember that, it can serve the dream. If it helps us forget it, the dream will not remain a dream for very long.

Closing Reflection: The Dream Is Conditional

Part Two began with hope, because fear alone cannot guide a future worth having, and the dream is real. Artificial intelligence can help humanity cure disease, expand access, educate more personally, support people with disabilities, reduce waste, improve agriculture, protect resources, and give people more time for the things that make a life worth living in the first place.

But the dream is not automatic. It requires a moral direction, requiring human beings willing to say that power is not the same as purpose, that speed is not the same as wisdom, that convenience is not the same as flourishing, that profit is not the same as service, and that control is not the same as safety, however much these pairs are so often confused for one another. If AI is going to help us become better, we first have to become clear about what better actually means, and that clarity will never come from the machine. It has to come from us, and it is ours to write, one honest choice at a time, whether we feel ready for the task or not.

PART THREE

The Divide

The danger is not only that AI becomes too powerful. The danger is that too few people control the power.

Chapter 7: The New Power Structure

The future will not only belong to the people who understand AI. It may belong to the people who own the systems everyone else depends on, and the dream is beautiful, deserving to be beautiful, but dreams become dangerous when they ignore power entirely. If Part Two asked what artificial intelligence could help humanity become, Part Three must ask who controls the road that takes us there, since the future will not be shaped by possibility alone. It will be shaped by ownership, by access, by capital, infrastructure, data, law, labor, and politics, and by the quiet decisions made inside companies and governments that most citizens will never see or even know to look for. The question is not only what AI can do. The question is who controls what AI can do, and that is where the divide begins to take shape.

The great promise of AI is that intelligence becomes more available than ever before in human history. A small business owner can use tools that once required an entire team of specialists. A student can ask for help at midnight when no tutor is available, and a farmer can analyze conditions faster, a researcher can sort through information with real assistance, and a person with a disability can gain access to tools that make daily life easier than it has ever been. That is real, and it matters more than we sometimes acknowledge, but there is another truth sitting beside it that we cannot ignore. The most powerful AI systems require resources most people simply do not have, enormous amounts of data, specialized chips, data centers, energy, highly trained technical workers, cloud infrastructure, and money at a scale that puts the real race in the hands of a small number of companies, nations, investors, and institutions who answer to almost no one but themselves.

The tool may feel democratic at the screen, but the system behind the screen is rarely democratic at all, and that distinction matters more than it first appears to the person simply typing a question into a box. A person using a free or low cost AI tool may feel they have access to the future, and in one sense, they do, but access is not the same as ownership, use is not the same as control, and dependence is not the same as power, however similar these things may feel from the outside. If five companies control the models, three companies control the cloud, two countries control most of the advanced infrastructure, and a narrow group controls the chips, data, standards, and distribution, then the ordinary user holds a powerful tool in one hand and very little say over the world being quietly built around them with each passing year.

Is that a reason to reject AI? It is not. It is a reason to stop pretending that access alone solves the ethical problem in front of us. The Stanford AI Index has tracked the rapid growth of private investment, adoption, and governance pressure around AI, and its most recent report points toward a world where the systems are moving quickly while the institutions around them struggle to keep pace with what they have built. The United Nations scientific panel has also warned that uneven infrastructure, uneven governance capacity, and uneven access could leave countries and communities dependent on systems they do not fully control and cannot meaningfully question.

That is the new power structure, not only rich and poor, not only employer and worker, not only government and citizen, but model owners and model users, data holders and data subjects, cloud owners and cloud renters, countries with compute and countries without it, companies with enough capital to build foundational systems and companies forced to build atop someone else's foundation entirely, and people who set the rules alongside people who only click agree without ever reading what they have agreed to.

Is that a small shift? It is not, since when intelligence becomes infrastructure, dependency becomes political in a way most people never intended it to be. We understand this already with roads, electricity, banking, water, food, and housing, since whoever controls the infrastructure controls more than a service. They control the conditions under which other people live and operate every single day, and AI may become that kind of infrastructure too, not visible like a bridge or a power line, but sitting quietly underneath decisions we rarely notice being made. It may sit underneath hiring, lending, insurance, policing, health care, education, search, and public benefits, underneath what people see, what people believe, what people are offered, what people are denied, and what people are told is possible for them to become.

That is why the divide is not only economic. It is moral, since a society can give people access to a tool while still denying them power over the systems that shape their lives from behind the scenes. A tenant may use AI to write a letter to a landlord, while the landlord uses AI to screen thousands of applicants at once without ever reading a single name closely. A job seeker may use AI to polish a resume, while an employer uses AI to filter applicants before a person ever reads their name at all, and a patient may use AI to understand symptoms, while an insurance company uses AI to evaluate the claim on entirely different terms. A citizen may use AI to fill out a form, while a government agency uses automated systems to flag fraud, assign risk, or deny benefits behind a wall of code no one outside can see. In each case, both sides are using technology, but they are not using it from the same position, and that difference in position is the one we cannot afford to miss.

Ethics has to care about position. It has to care about who holds power and who is being acted upon, who can appeal, who can ask why, who can afford help, who can leave the system, and who has no choice but to live under it regardless of how they feel about the arrangement. The same technology feels entirely different depending on where you happen to stand. To the company, AI may feel like efficiency. To the worker, it may feel like surveillance. To the bank, it may feel like risk management, and to the applicant, it may feel like rejection without explanation of any kind.

This is one of the oldest ethical problems wearing a new form, since power always finds a way to name itself as order. The person with power says this will make things more efficient. The person under power says this may make me invisible. The person with power says the system is objective. The person under power says I cannot challenge it, and that is why we must be careful with the word objective, since AI can create the feeling of objectivity simply because the decision comes from a system rather than a person standing before us. It can make a decision feel clean because the mess of human judgment hides behind technical language most people were never trained to question, but technical language does not remove human choice. Someone chose the goal, the data, the model, the deployment setting, the level of review, the business model, and what would count as success in the first place, and a system is never free from values. It only hides them better than we do ourselves.

The AI Now Institute has repeatedly argued that the AI debate is also a power debate, especially where large technology firms shape markets, public policy, and the direction of development in ways few outside the industry ever get to observe closely. That point belongs here, because ethics without power analysis becomes too polite, speaking of principles without asking who benefits when those principles are quietly ignored behind closed doors. We do not need to hate business to ask business hard questions, and we do not need to distrust every government to ask government hard questions either, but we do need to ask, plainly and without apology, who owns the tool and who answers when it harms people who never had a say in its design.

The divide grows whenever benefits travel upward and risk travels downward, and that sentence should haunt us a little longer than it usually does. The executive receives efficiency. The worker receives pressure. The platform receives data. The user receives dependency, and the government receives speed while the citizen receives less explanation than before. Is this inevitable? It is not, but it is possible, and because it is possible, it deserves our serious and sustained attention rather than a passing mention in a policy paper.

A hopeful AI future cannot be built on a structure where a few people own the intelligence layer of civilization while everyone else merely rents access to it on terms they did not negotiate. A humane AI future requires shared benefit, public understanding, contestability, open standards where possible, real competition, regulation where rights are at stake, and a genuine voice for workers, citizens, and small businesses who are so often left out of the rooms where these decisions get made. It requires poorer countries not to become permanent customers of tools built entirely around someone else's language, values, data, and priorities, a point that matters more than many people realize, since AI systems trained largely in dominant languages and built around dominant markets may not understand the cultural, economic, legal, or social realities of smaller communities at all.

A model can appear universal while still carrying the assumptions of those who built it, and this is not always malicious. Sometimes it is simply the result of where money, data, and talent happen to be concentrated, but harm does not need hatred to become real. Neglect is often enough on its own to do lasting damage, and a world divided by AI may not look like science fiction at all. It may look like ordinary life, only more automated than before, where some people receive customized medicine, personalized education, AI supported legal help, and powerful productivity tools, while others receive automated rejections, suspicion scores, underfunded services, and systems that do not speak their language well enough to understand their lives in any meaningful sense.

That is the future we have to resist, and the dividing line may not be between people who use AI and people who do not. It may be between people who are helped by AI and people who are merely managed by it, processed rather than served, counted rather than known. That is the warning of this part of the book. The dream is possible. The divide is also possible, and if we do not name it clearly now, while the page is still open before us, we will build it quietly anyway, one convenient decision at a time, until one day we look up and find it already finished.

Chapter 8: Business, Profit, and the Moral Temptation

Efficiency is not the same as service. Profit is not the same as purpose. Speed is not the same as wisdom, and business will be one of the first places most people feel AI, not because business is evil, but because business moves quickly whenever something saves time, reduces cost, opens capacity, creates leverage, or increases profit for the people at the top of it.

Is that automatically wrong? It is not, since a business should look for better ways to serve, should eliminate waste, use tools that help employees do better work, reduce errors, communicate more clearly, decide with better information, and improve the customer experience wherever it can. AI can help with all of that, and a small business owner can use it to write clearer emails, summarize documents, prepare marketing, build training materials, analyze feedback, organize processes, draft job descriptions, compare vendors, and read contracts more carefully than time alone would ever allow. I believe that is good, plainly and without reservation.

But the moral temptation in business is always the same one, regardless of the industry or the size of the company involved. The tool that begins as a way to serve people can become a way to squeeze them instead, quietly, a little at a time, until the squeezing feels like standard practice rather than a choice anyone made. That is where ethics has to speak clearly, before the drift becomes permanent.

Business does not become unethical because it makes money. Profit is not the enemy here. Profit keeps the doors open, pays people, creates stability, and allows a company to grow, innovate, hire, and continue serving its customers for years to come. The problem is not profit. The problem is profit without restraint, the moment customers become targets rather than people, employees become cost centers before they are treated as human beings, speed becomes more important than truth, and a company hides behind AI because it wants the benefit of a decision without the burden of responsibility for having made it.

The company says the system decided, but does a system absolve the company of anything at all. It does not, since a company selected the system, trained it or purchased it, deployed it, chose where to use it, chose how much review to require, chose what level of error was acceptable, and chose whether customers could appeal at all when something went wrong. There is no moral escape hatch called software, however convenient that phrase might sound in a meeting.

This matters enormously in marketing, where AI has created a flood of new claims. Tools promise effortless growth, automatic income, instant expertise, perfect content, better decisions, safer screening, smarter sales, and lower risk, and some claims are true, some are exaggerated, and some are nonsense wearing a futuristic suit that fools no one who looks closely. The Federal Trade Commission has warned businesses to keep AI claims accurate and has taken action against deceptive AI related claims and schemes that promised far more than they delivered. Should that surprise anyone. It should not. The ethical rule is simple: do not use the excitement around AI to sell people something that fails to do what you say it does, however tempting the shortcut may seem in the moment?

That sounds basic, because it is basic, and most ethical failures begin with basic things people convinced themselves were complicated enough to excuse. Tell the truth. Do not mislead people? Do not promise certainty where there is only probability, and do not sell automation as expertise when a qualified human being remains genuinely necessary? Do not pretend a tool is safe merely because it is impressive to watch in a demonstration? AI does not remove the duty of honesty. It raises it, and raises the stakes for anyone tempted to look away from that duty.

This matters just as much in privacy, since many AI systems depend on information, some public, some from customers, employees, vendors, or private conversations, documents, images, and recordings that people never fully understood would become fuel for machine learning or automated decision making somewhere down the line. Privacy cannot be treated like a buried setting tucked away where no one will look. A person should not need to be a lawyer to understand what is happening with their own data, and a customer who shares information does not automatically consent to every future use someone later discovers a clever way to extract from it. An employee who writes messages at work does not become raw material for unlimited surveillance simply because the technology now makes that surveillance possible.

The FTC has specifically addressed privacy and confidentiality commitments in the AI context, including the risk that companies may benefit from models built with data obtained in ways that violate law or promise. Is that a big deal. It is, because it reminds us that data is not just data. Data came from people, and behind every data set there are lives, faces, names, habits, mistakes, preferences, fears, and histories people may not even know they are revealing to a system they have never met.

Business ethics has to begin with that recognition and hold onto it firmly. A person is not a data mine. A customer is not a pile of signals, and a worker is not merely a productivity graph updated in real time. A business using AI ethically should be able to answer some plain questions out loud, without flinching. Are we using this to serve people better? Are we honest about what the system can and cannot do? Are we protecting the information people trusted us with, and can a human review important decisions, can a person actually appeal when something goes wrong? These questions are not anti business. They are good business, because trust is not a soft thing. Trust is an asset, the very thing that allows customers to share information, employees to speak honestly, and communities to support a company through hard seasons rather than abandoning it the first chance they get.

Can AI create short term efficiency while destroying long term trust? It can, easily and often without anyone noticing until the damage is done. A company can automate too much of the customer experience and call it innovation while customers experience it as abandonment, plain and simple. A company can replace human support with a chatbot and celebrate the savings while people sit in loops, unable to solve real problems that a single phone call once resolved in minutes. That is not service. That is distance, and business leaders should be careful about distance, since it becomes easier to ignore the human effect of a decision once no one has to look another person in the eye and explain it. Not because the leader is evil, but because abstraction makes harm easier to overlook, day after day, until it no longer registers at all.

This is why accountability matters so much here. The NIST AI Risk Management Framework is useful, because it pushes organizations to think about AI risks in a structured way rather than treating risk as a public relations issue only after something has already gone wrong and the damage is public. The OECD AI Incidents Monitor reflects the same growing need to learn from real AI incidents rather than pretending harm is only theoretical until it happens to someone else. A serious business cannot say we use AI but do not know the risk. That is not leadership. That is negligence with better branding attached to it.

Hiring offers one of the clearest examples of where this goes wrong. AI tools may help sort applicants, match skills, create interview questions, and identify patterns in candidate pools, and used carefully, they may reduce administrative burden considerably. Used carelessly, they may screen out qualified people, reproduce old bias, punish gaps in employment, or misread disability related differences, creating a hiring process nobody inside the company can even explain when challenged on it. The EEOC has treated AI and algorithmic fairness in employment as a serious area of concern, and multiple United States agencies have stated plainly that automated systems must still comply with existing laws around discrimination, consumer protection, fair lending, and civil rights regardless of the technology involved. The law does not disappear because a company used a model. Neither does morality, however the paperwork is worded.

A business should never use AI to do at scale what it would be ashamed to do to one person at a time, and that may be the single most useful test in this chapter. Would we say it directly to a customer? Explain it directly to an employee. Defend it in public, allow the people affected to understand how the system works, and use the same system on our own family without hesitation. If the honest answer is no, the company should slow down immediately, whatever the quarterly numbers demand.

The moral temptation of AI in business is not only that it can do bad things. The deeper temptation is that it can make bad things feel entirely normal over time, more tracking, more scoring, more filtering, more nudging, more replacing, and more pretending that because something is possible, it must therefore be acceptable. That is the line leaders must refuse to cross, no matter how much competitive pressure pushes them toward it.

AI should not become a machine that extracts more from people while calling the extraction service. It should help businesses serve with more clarity, more competence, more truth, and more care than they managed before these tools existed. A business that uses AI well should become more human in the places that matter most, not less, and that possibility remains entirely open, waiting on the choices business leaders make today, and tomorrow, and the day after that, one decision at a time.

Chapter 9: Government, Surveillance, and the Machinery of Authority

When government uses AI, the question is not only whether the system works? The question is what happens to the citizen when it does not, since government holds a different kind of power than business ever does. A customer can sometimes leave a company, but a citizen cannot leave the state so easily, which is exactly why government use of AI requires a higher standard than any private use could reasonably be held to.

When a business uses AI poorly, people can be misled, exploited, denied, watched, or manipulated, and that is serious? But when government uses AI poorly, the consequences can touch freedom, benefits, policing, education, immigration, courts, public housing, child welfare, health programs, taxes, licenses, and the basic relationship between a person and authority itself, a relationship that cannot easily be renegotiated once it has soured. Government does not simply offer services. Government can compel, investigate, punish, deny, remove, separate, and take, and because of that, government should never be allowed to hide behind a system when these powers are exercised.

The phrase the algorithm decided should never exist in a free society as a final answer to anyone. A citizen deserves a person, a reason, and a path to challenge the decision made about their life. A citizen deserves to know whether an automated system affected the outcome, and deserves more than a score standing in for an explanation that a human being could actually give.

Is every government AI tool bad? That would be far too simple, and it is not true at all. There are good uses, and AI may help cities manage traffic, detect infrastructure problems, translate public information, improve emergency response, identify waste, process routine paperwork, and help agencies serve people faster than they otherwise could. Public systems are often old, slow, understaffed, and frustrating, and there is real room for improvement in how they function. The ethical question is not whether government can use AI. The ethical question is where, how, why, under whose authority, with what oversight, and with what rights reserved for the people affected by the decision.

A chatbot answering basic city hall questions is one thing. A system deciding who gets investigated for fraud is another entirely, and a translation tool on a public website is one thing, while a predictive policing system is another altogether. Different uses carry different moral weight, which is why risk-based approaches matter so much here. The European Union AI Act treats some AI uses as higher risk because they affect sensitive areas such as biometrics, critical infrastructure, education, employment, essential services, law enforcement, migration, and justice. One need not agree with every detail of any particular law to recognize the moral instinct behind that structure, which is simply this: a system that recommends a movie should not be judged the same way as a system that recommends whether a person is dangerous. The higher the stakes, the higher the burden ought to be, and that should be obvious, though efficiency has a way of making obvious things negotiable when no one is paying close attention.

Government loves efficiency for the same reason business does. It saves time and money, helps staff do more with less, creates dashboards, and gives leaders the feeling that a complex public system can finally be organized and understood at a glance. But is public life only a management problem. It is not, since a government can become very efficient at being unfair, at denying help, at watching people, and at processing poverty instead of reducing it, all while the dashboards report success. That is why public sector AI must be judged by more than speed and cost savings alone.

AlgorithmWatch has studied automated decision-making systems in public administration and emphasized the importance of ethical evaluation, transparency, and accountability for those affected by these systems. Virginia Eubanks brought broader public attention to how automated systems can profile, police, and punish poor people through public service systems that were meant to help them, and Cathy O'Neil helped popularize the danger of opaque, scalable algorithmic systems that can affect people without meaningful recourse of any kind. The common thread is not that every system fails. The common thread is that when systems do fail in public life, they most often fail against people with the least power to fight back against the failure.

Should that concern us? It should, since a wealthy person can hire a lawyer while a poor person may get a letter they do not understand and have no one to help them decode. This is the moral danger of automated authority. It can make power feel less personal while making it harder to challenge at exactly the same time, a combination that should trouble anyone who values a free society.

Surveillance adds another layer entirely to this problem. Government has always wanted to see more, and that is not new. Census records, tax records, police reports, school records, welfare records, and property records all existed before AI arrived on the scene, but what changes with AI is what can be done with that information once it is gathered. More data can be connected, more patterns inferred, more people monitored, more behavior predicted, and more suspicion automated before a human being ever pauses to ask whether the underlying premise was wrong to begin with.

The deeper concern is not only cameras, facial recognition, or location data, though those matter too. It is the construction of a public environment where ordinary life becomes constantly legible to systems of authority, where you go, who you know, what you buy, what you search, what neighborhoods you enter, and what patterns your life appears to fit according to a model that has never met you. Can that become a quiet kind of control? It can, since people behave differently when they believe they are always being watched. They speak differently, gather differently, and ask for help differently than they would in a society that trusted them by default. A free society cannot be built on constant suspicion of its own citizens.

Is transparency alone enough? It is not, since a government cannot simply say we disclosed the system and call the matter settled, because a citizen may know a system exists and still have no power over it whatsoever, no way to see inside it or challenge what it decides. Transparency without contestability is a window that does not open, and accountability has to include the ability to act, not merely the ability to observe from a distance. In government, the citizen must be able to appeal. The decision must be reviewable by a human being with real authority to change it, and the agency must be able to explain the reason in plain language rather than technical jargon designed to discourage questions. Harmful systems must be capable of being paused or removed entirely when they fail, and if not, transparency is theater rather than substance.

The United Nations panel has warned that AI capabilities and risks may be moving faster than governance and scientific understanding in some areas, a warning that matters especially for government, since public systems tend to be slow to change once adopted. A bad tool can become embedded in procurement, training, budgets, and institutional habit before anyone notices what has happened, and once a system is inside government, it may become harder to remove than it was to purchase in the first place, a fact that should give every agency pause before adoption rather than after.

That is why public agencies need humility above almost anything else. Do they have the expertise to evaluate what they are buying? Have vendor claims been independently tested by anyone outside the sales process? Have affected communities been consulted before deployment, and will the system reduce harm, or simply move harm out of sight where it becomes harder to measure? Citizens are not beta testers for authority, and a public AI system should never be deployed against people first and understood later, which is exactly backward from how responsible governance ought to work. Government must move carefully wherever rights are at stake, testing, auditing, disclosing, preserving human review, and limiting surveillance rather than expanding it simply because the technology now makes expansion easy.

None of this means government must fear every new tool that comes along. It means government must become worthy of public trust, and trust is not created by telling people to trust the system. Trust is created when the system can be questioned, when errors can be corrected, when officials remain responsible for what happens under their watch, and when the poor are not experimented on simply because they have fewer ways to object to what is being done to them.

The machinery of authority will always be tempted to make people easier to manage than to serve, and ethical AI in government must resist that temptation at every turn, because the purpose of government is not to automate control. It is to serve justice, order, safety, liberty, and the common good, all of which require a human hand remaining on the wheel. If AI helps government do that with more fairness and more humanity, it deserves consideration, but if it helps government avoid explanation, expand surveillance, or hide responsibility behind a wall of code, it deserves resistance instead, from citizens and officials alike.

The citizen is not the subject of the machine. The machine must remain subject to the people, and whether it does or not remains, even now, an open question we are still answering through every choice a government makes about the tools it adopts.

Chapter 10: The Labor Question

When work changes, people do not only lose tasks? They can lose rhythm, identity, status, and the place where they once knew they mattered to someone beyond themselves, and the labor question is not only whether AI will take jobs. That is the surface question. The deeper question is what happens to people when the economy no longer needs them in quite the same way it once did, and whether we have prepared ourselves for the answer.

Is work only income? It is not, since work is rhythm, identity, competence, status, and contribution all at once, the place where many people learn discipline, patience, teamwork, responsibility, pride, and sacrifice, where they become useful to others and are known by them in return. So when we speak of AI and work, we must be careful, because we are not only speaking of tasks. We are speaking of people, and of everything a task has quietly come to mean to the person performing it.

The economic reports matter here, since the IMF has estimated that AI may affect a large share of global employment, replacing some work and complementing other work, with the possibility of productivity growth alongside the risk of deeper inequality taking hold. The International Labour Organization has studied occupational exposure to generative AI, focusing on the difference between automation and augmentation across tasks, and the Stanford AI Index tracks emerging evidence about labor market effects and the pace of adoption across industries. Those reports matter, because they give structure to the conversation, but the lived reality will be felt in far more ordinary places than any report can capture. A legal assistant wonders whether the first draft will still be her own work. A customer service representative watches the chatbot handle more calls each week. A graphic designer watches clients turn to image generators before ever calling a professional, and a writer wonders whether people still value a human voice at all in a world so full of generated ones.

People are not afraid of innovation in the abstract. People are afraid of being made unnecessary, and that fear deserves respect, not panic, but respect, since the person who says AI will simply create new jobs may be right over a long enough timeline and still fail entirely to comfort the person whose job disappears next month with no warning at all.

History tells us that technology changes work, replacing some tasks, creating others, raising productivity, disrupting old industries, and creating new skills while destroying old certainties that once felt permanent. That pattern is not new, but AI feels different because it reaches into cognitive work in a way earlier machines never did. It does not only lift, move, cut, or calculate. It writes, summarizes, translates, codes, drafts, designs, analyzes, and plans, touching work that many people believed was protected because it required education, language, creativity, or professional judgment that no machine could replicate.

For decades, many people were told to get educated so that automation would not reach them, and now the machine can write emails, draft contracts, generate code, and imitate the surface of expertise convincingly enough to fool a casual reader. Does that mean it truly understands any of it? It does not, but in a labor market, the mere appearance of competence can still change hiring decisions in ways that matter a great deal to the person affected. A company does not have to believe AI is perfect in order to reduce headcount. It only has to believe the savings are worth the mistakes that will inevitably follow, and that sentence matters more than it first appears, because the ethical problem is not only whether AI can do the work as well as a person. The ethical problem is whether leaders will accept lower quality and greater social cost simply because the financial model rewards them for doing so quietly, without anyone having to say it aloud.

This is where business ethics and labor ethics meet directly, without room for evasion. If AI helps a worker become more capable, that is one thing entirely. If AI turns the worker into a supervised extension of the machine, that is another, and the future of work cannot be measured by productivity alone, because a society can become more productive and less humane at exactly the same time, producing more output while people feel less secure, generating more content while human creativity is valued less, and responding faster while workers become more replaceable with each passing quarter. Is that progress? Only if we define progress very poorly indeed, mistaking the numbers for the whole story.

A humane labor future has to answer two questions rather than one, since people need to know how they will earn a living, and they need to know that they still matter beyond their paycheck. The first question is economic. The second is spiritual, cultural, and social, and we often talk about the first while quietly ignoring the second altogether. That is a mistake, because income matters deeply. People need to pay rent, buy food, care for children, and live with real stability, and any serious AI economy has to address wages, retraining, worker bargaining power, and the broader social safety net that catches people when a transition goes wrong. But does money alone solve meaning. It does not, since a person can receive assistance and still feel useless, can have more free time and still feel lost, still wonder what his life is for once the tasks that once defined him have quietly disappeared.

That is why the utopian dream from Part Two must be handled with real care here. A world where people no longer have to do meaningless work could be beautiful, but a world where people no longer feel needed could become dangerous in an entirely different way, one this book has already warned against. Human beings need purpose, and if AI removes tasks, we must rebuild pathways to contribution rather than simply celebrating the removal. If AI shortens work, we must strengthen family, art, service, learning, faith, community, and civic life to fill what has been opened up. If AI creates wealth, we must ask who shares in it, and whether the gains flow only to shareholders and executives or reach the workers and communities who helped create them in the first place.

That is the real labor question: who receives the gain, and who carries the loss? If the answer stays the same every time, the divide will only grow wider with each passing cycle of innovation.

Leaders should be honest about this, because too many conversations about AI and work hide behind soft language. We say streamline, optimize, right size, and unlock efficiency, and sometimes that language is true, but sometimes it simply means fewer people, and workers know the difference even when they cannot name the technical details behind the decision. They know when a tool is introduced to help them, and they know, just as surely, when it is introduced to replace them entirely.

A company that wants to use AI ethically should bring workers into the conversation early, tell the truth, train people, give them time to adapt, and share the benefits when productivity rises rather than keeping them for a select few at the top. It should not ask employees to train the very tool that will quietly eliminate their position without acknowledging the moral weight of that request, however common the practice has become. That is not sentimental. It is leadership, and the labor market is not made of abstractions. It is made of fathers, mothers, sons, daughters, neighbors, veterans, and immigrants, people starting over and people trying to prove they still have something to give the world. We should speak of disruption with that level of care every single time, not the language of a quarterly earnings call.

There is another side to the labor question as well, and it deserves to be named honestly alongside the warnings. AI may make more people capable of doing things they could not do before. A small team may operate with the capacity of a much larger company. A person without formal training may learn faster than before, and a rural business may access expertise it never could have afforded on its own. A disabled worker may gain new independence, and a parent may start a business from home that would have been impossible a decade ago. That is not minor. That is part of the promise, and the goal is not to freeze work in place forever but to make change humane wherever it occurs. We should not protect every old task, but we should protect people, and that distinction matters more than almost anything else in this chapter.

Training is part of the bridge that carries people through change without losing them along the way. So is time, shared benefit, honesty, worker voice, and a culture that values people beyond their economic output alone. Without that bridge, AI becomes another force telling people to adapt while the rewards flow somewhere else entirely, out of reach, and that is how division hardens, not all at once, but one decision at a time, one department at a time, one family at a time, until the pattern becomes too familiar to notice.

The ethical labor future is not anti AI. It is pro human, saying the worker is not an obstacle to efficiency but a person whose dignity must travel with us into whatever comes next. If AI can help us build more, it should also help us build better lives alongside it. If AI can increase productivity, it should also increase human possibility, and if it can create wealth, we must have the courage to ask whether that wealth serves the common good or only climbs upward, away from the people who made it possible in the first place through their labor.

The labor question will test our ethics, because it will test whether we mean what we say about human dignity when the savings on the spreadsheet are real and immediate. It is easy to say people matter in a chapter about philosophy. It is far harder to say people matter when a spreadsheet insists the company can save twenty percent by removing them. That is when ethics becomes real, and it is also, if we choose it, when hope becomes real too, in the choices we make when no one is forcing our hand.

Closing Reflection: The Divide Is a Choice

The divide does not arrive with a dramatic announcement. It arrives through ordinary decisions, a company choosing efficiency without transparency, a government choosing automation without appeal, a school choosing detection without trust. That is how the future bends, not all at once, but quietly, decision by decision, until the shape of things has changed without anyone quite deciding it should.

This part of the book matters because doom gets attention for being dramatic, and the dream gets attention for being beautiful, but the divide may be the most likely danger of all, since it requires nothing so extraordinary as a machine waking up to hate us. It only requires human beings to do what human beings have so often done with powerful tools before: concentrate the benefit, distribute the cost, and call it progress.

But is any of that fixed in place. It is not. AI could help cure disease, educate children, strengthen communities, support small businesses, improve public services, and give people more room to become fully human, or it could become a system where the powerful are assisted and the vulnerable are merely processed like inventory. The difference will not be decided by technology alone. It will be decided by ethics, by ownership, by law, by leadership, and by whether ordinary people have the right to question systems that affect their lives directly.

The question is not whether business should use AI. Business will use it. The question is whether business will use it to serve or to extract from the people it claims to serve. The question is not whether government should modernize. It must. The question is whether modernization will strengthen public trust or hide public power behind machines that answer to no one. The question is not whether work will change. It will. The question is whether people will be invited into that change with dignity, or forced to absorb it alone, quietly, without a bridge to stand on.

That is the divide. It is not only technical. It is human, and because it is human, the choosing is still ours to make, today and every day after this one.

Part Four

The Personal Machine

The most intimate ethical test may not be what AI does in public. It may be what AI does in the places where we are most human.

Chapter 11: AI in the Home

The ethical debate does not stay in the laboratory. It follows us into the kitchen, the bedroom, the car, the phone, and the quiet moments when no one else is watching us at all, and artificial intelligence will not only arrive through governments, boardrooms, factories, courts, and schools. It will arrive through the front door, living in the phone on the nightstand, the search bar, the calendar, the car, the doorbell, the watch, the thermostat, the television, and the device that listens when someone speaks its name from across the room without a second thought.

For most people, will AI feel like a revolution at first. It will not. It will feel like convenience, writing the grocery list, summarizing an email, reminding someone to take medicine, telling a driver which way to go, and it will be useful, which is precisely what makes it powerful in ways we rarely stop to consider. Most technology does not take over daily life by force. It takes over daily life by becoming helpful, saving a minute, then an hour, until the routine becomes dependence and the dependence becomes normal without anyone quite noticing the transition.

Is that always bad? It is not, since we depend on electricity, roads, clean water, and grocery stores, and human life has always been built on layers of trust in systems we did not build ourselves and could not fully explain if asked. The real question is not whether dependence is wrong. The question is whether we understand what we are depending on, who controls it, and what it is doing while it helps us with the small tasks of an ordinary day.

That is the ethical question inside the home. A tool that helps with dinner may also be collecting patterns about a family without their full awareness. A tool that helps with homework may also be shaping a child's confidence in ways no one intended, and a tool that helps with security may be building a private record of everyone who comes and goes through that door. The home used to be the place where people escaped many of the systems that measured them elsewhere. It was never perfect, but there was a line between public life and private life, a difference between the office and the kitchen table that everyone understood without needing to say it aloud.

Does AI weaken those lines? It does, not because every company is trying to harm people, but because the modern home is becoming a field of sensors, prompts, predictions, microphones, cameras, accounts, and data flows that most families never signed up to host. A family may not think they are living with AI at all. They may simply think they bought a smart appliance, downloaded an app, or accepted the default setting because the alternative seemed inconvenient. But is free ever really free. Rarely, and sometimes the price is attention, sometimes it is privacy, and sometimes it is the slow surrender of decisions that used to belong entirely to the family gathered around the table.

Pew Research Center's work on public views of AI and data privacy shows that ordinary people are not simply excited or simply afraid. Many are trying to understand how these systems will affect society, work, information, privacy, and personal life, often without the language to fully articulate their unease. That matters, because the personal machine enters daily life before most people have been given the vocabulary to think about it clearly, let alone push back against it where pushing back would help.

If a family only calls a tool convenient, will they ever stop to ask whether it is shaping them in return. Probably not, and if a parent only calls a tool educational, will they ask whether it is quietly weakening attention instead. Perhaps not, and the word smart should not end the conversation. It should begin one, asking who the tool is smart for, what purpose it serves, whose data makes it work, and who controls it once the data has left the house.

That last question matters, because AI in the home is not only about privacy. It is also about authority, and the more useful a system becomes, the more authority we may quietly hand over to it without ever making a conscious decision to do so. It suggests a route, and we follow it. It suggests a product, and we buy it. It suggests a reply, and we send it without pausing. At some point, does the tool even need to command us. It does not. It only needs to suggest, and a suggestion, though softer than an order, can still change a life over enough repetitions.

This is why personal AI has to be understood as more than a gadget sitting on a counter. It is becoming a quiet decision layer in ordinary life, and it does not only answer questions. It frames them. It does not only respond to preference. It trains it, shaping what we notice and what we come to expect, since a person is formed by what they repeatedly see, just as a family is formed by what it repeatedly gives its attention to over months and years.

The home will need boundaries, not panic, but boundaries drawn with real intention. A healthy family does not need to reject every tool, which would be neither possible nor wise given how thoroughly these tools have woven themselves into daily life. AI will be useful for planning, learning, translation, safety, accessibility, and much of daily life besides. But a healthy family also has to know the difference between help and authority, using AI to make a grocery list while refusing to let AI decide what the family values, using AI to explain a medical term while refusing to let it replace a qualified doctor, using AI to help a child practice a concept while refusing to let it become the child's primary teacher, companion, or judge of worth.

This is where ethics becomes practical rather than abstract, since it is easy to debate AI in the abstract and far harder to decide what happens at the dinner table on an ordinary Tuesday evening. For families, the boundaries need to be concrete. Phones may need to stay out of bedrooms. Children may need supervised access to chatbots rather than open, unmonitored use. Families may need a verification phrase, since a realistic voice on the phone may not belong to the person it sounds like at all, however convincing the imitation.

The Federal Trade Commission has warned about harmful voice cloning and has treated it as a consumer protection issue rather than a merely technical curiosity confined to research papers. That is exactly the kind of risk that proves the ethical debate belongs inside the home, because a fake voice does not need to fool a government agency to cause real harm. It only needs to fool a parent, a grandparent, or a frightened person in a vulnerable moment when clear thinking is hardest to summon.

The most dangerous AI uses may not always look dramatic. They may sound like someone you love, and that changes family life in ways we are only beginning to reckon with. It means trust now needs practices, call back habits, and rules about money requests that would have seemed paranoid a decade ago. It means teaching children and older relatives that urgency is so often part of manipulation rather than a sign of genuine emergency. In a world of generated voices, can trust depend only on recognition. It cannot. It has to depend on verification, and that is not paranoia. That is wisdom, plainly stated and worth repeating to everyone in the house.

Connected devices add another layer of concern to all of this. The Internet of Things was already placing sensors and software into the home before generative AI became part of everyday conversation, and the FTC has published guidance on securing connected devices as consumer reporting continues to raise privacy questions around smart appliances that quietly gather far more than they need. As AI becomes embedded in these devices, does the privacy question grow larger. It does, and a refrigerator that tracks groceries may be helpful, while a car that records driving behavior may improve safety, but does every one of those examples also create information about a life. It does, and a home can reveal more than people realize, learning when a family wakes, who is home, who is sick, who is anxious, and who is struggling with money in ways they never intended to disclose to a machine.

If that information has value, will someone want it. Someone always will, and the ethical question is whether people have meaningful control over the digital traces of their own lives, whether they can see what is collected, delete it, and use the device without being pushed into another subscription or another surrender of the very privacy they were told the product would protect.

A household cannot be expected to become a full-time cybersecurity department, yet ordinary people are increasingly asked to manage passwords, devices, identity theft, scams, and now AI systems capable of generating convincing text, images, audio, and video on demand. Consumer Reports, Aspen Digital, and the Global Cyber Alliance have treated consumer cyber readiness as a public issue, because personal security has become part of ordinary life whether we asked for that responsibility or not. That is the burden we have placed on families, and it is too much to put entirely on the user standing alone at the kitchen counter. Ethical AI in the home cannot mean families are responsible for protecting themselves from every risk while companies collect the benefit of the data flowing the other direction. The burden has to move upstream, where design has to become safer, defaults have to become more protective, and consent has to become clearer than a checkbox nobody reads.

The household deserves respect. The home is not just a market. The family is not just a user base, and the child is not just a future customer waiting to be cultivated. The kitchen table is not just another data source, since the home is where people become who they are, and AI entering that space has to carry a higher duty because of what is at stake there.

This chapter is not an argument against personal AI. It is an argument for personal AI that understands the sacredness of the personal, since there are parts of life that should not be optimized to death, conversations that should not be recorded, and moments that should not become data points in someone else's model. There are children who should not have their development shaped by systems designed primarily to hold attention, and there are lonely people who should not be captured by a system that learns exactly what to say in order to keep them coming back.

That is the personal machine. It is not always dangerous, and it is not always harmless. It is powerful because it is intimate, and the ethical question here is simple enough to say and hard enough to live consistently. The tool should help the people inside the home become more present, more capable, more connected, and more free, and if it trains them instead to become more dependent, more watched, more distracted, and more alone, then the convenience is costing far too much, a price we still, in every home, get to decide whether we are willing to pay.

Chapter 12: Children, Education, and the Formation of the Mind

The goal of education is not answer production. The goal is the formation of a mind capable of judgment, imagination, discipline, and care, and the first public argument about AI in education was predictable enough: are students cheating? That question matters, but is it large enough to hold the whole issue before us? It is not, since cheating is a real concern that schools, teachers, and parents all have to deal with directly, but if cheating becomes the center of the conversation, we will miss something far larger unfolding underneath it. AI is not only changing how students complete assignments. It may be changing how students form their minds in the first place, and that is a much more serious matter than any single assignment.

Does a child become educated because answers appear quickly? No child ever has, since a child becomes educated through attention, struggle, memory, correction, curiosity, effort, repetition, conversation, failure, and the slow building of judgment across years that cannot be shortened without cost. Learning requires friction, not cruelty, not humiliation, not pointless busywork, but friction, the honest resistance a mind must push against in order to grow at all. A student has to sit with confusion long enough to develop patience, has to try a sentence, erase it, and rewrite it until the thought finally becomes their own rather than something borrowed.

If AI removes all friction, does it also remove formation along with it. It does, and that is the real danger here, since a student who always receives the polished answer may never develop the muscles that produce one on their own. A student who never struggles with writing may never learn the discipline of thought that writing itself creates in the struggling. Education is not a vending machine. You cannot insert a prompt and receive wisdom in return, however convincing the illusion may feel in the moment.

This is why AI in education has to be handled with both courage and caution held together at once. Should we pretend students will not use it? We should not, since they will regardless of any policy written against it. Should teachers try to ban their way back into the past? They cannot, and should every AI use be treated as cheating. That would be lazy, and should every AI use be treated as learning. That would be just as lazy in the opposite direction. The better question is this: what kind of use strengthens the student, and what kind of use weakens the student? That single question can guide almost everything that follows from here.

If AI helps a student understand a concept after honest effort has already been made, it may strengthen the student. If it replaces the honest effort entirely, it may weaken the student instead, quietly and over time. If it helps a student with a disability access material or express ideas more fully than they could alone, it may strengthen the student, but if it quietly lowers expectations because the system assumes the student cannot do more, it may weaken the student in ways no one intended or even noticed happening.

UNESCO's guidance on generative AI in education and research emphasizes a human centered approach, policy development, AI literacy, and the need to evaluate tools for educational and ethical appropriateness before they are adopted widely. The United States Department of Education has also framed AI in teaching and learning around the need for human judgment, policies, teacher support, and educational purpose above all else. That phrase, educational purpose, matters more than it first appears, since the purpose is not simply to produce an answer. The purpose is to form a person capable of producing their own.

Once we remember that, the role of the teacher becomes clearer than any policy document could make it. A teacher is not only a content delivery system. A teacher sees the student, hears the hesitation, and knows whether the silence in the room is confusion, fear, boredom, or hunger that has nothing to do with the lesson at all. Can a machine love a student? It cannot, and it can simulate patience, explain a concept, and generate practice questions, all of which are useful in their proper place, but is care the same as responsiveness. It is not. A teacher does more than answer. A teacher bears witness to growth over years, which is why the future of education should be framed as AI serving teachers, students, and families under clear human authority, never replacing any one of them.

That means children deserve special protection in this conversation. Children are not small adults. They are developing human beings whose judgment, identity, and ability to handle praise, rejection, and persuasive technology are all still forming and easily bent. Should the ethical burden be higher when AI systems interact with children directly? It should, and UNICEF's guidance on AI and children begins with children's rights, development, dignity, participation, privacy, fairness, and protection, which is the right place to begin any policy touching their lives. Children are not merely users, students, or data subjects to be optimized. They are people in formation, and that phrase deserves to stay with us through everything else this chapter has to say.

A child is allowed to surprise us, and that matters, because predictive systems can quietly become prisons for the very children they claim to help. A system may decide a child is weak in math, reads below level, or is disengaged from the material, and sometimes those signals genuinely help. But is a signal the same as a soul. It is not, since a child is more than the data collected at one point in time. A child changes, matures, has bad days, and has hidden strengths no assessment could capture in a single sitting. Ethical education technology must leave room for becoming, since that is one of the most human things about children. They are not finished, and neither, for that matter, are we as the adults responsible for them.

Education should not be reduced to workforce preparation alone. Work matters, and skills matter, but is a child born merely to become employable. No child is, since a child is born into the possibility of a full human life, and education should help that child learn how to think, speak, listen, create, question, cooperate, disagree, imagine, serve, and live well with others long after any test has been forgotten.

The question of writing deserves particular attention here, since writing is not only communication. Writing is thinking made visible, and when a person writes, they discover what they know, what they do not know, and what they are actually trying to say before the words arrived to say it. If a student uses AI to avoid writing altogether, do they lose more than an assignment. They do. They lose a chance to meet their own mind in the quiet space where that meeting usually happens. But if a student uses AI to ask for feedback, compare structure, or practice revision after producing their own work first, the tool may help the student become genuinely stronger rather than merely faster. The difference is rarely the tool itself. The difference is the posture behind its use, and the honest personal test is simply this: am I using the tool to escape thought, or to deepen it.

Students need AI literacy, though not merely learning how to prompt a system effectively. They need to learn how to question the answer, to understand that confidence is not proof, that generated text can be wrong, and that a machine can sound certain while making things up entirely without any awareness that it has done so. Most of all, AI literacy means learning that having access to an answer does not make a person wise. Wisdom still requires judgment, and judgment still has to be formed, slowly, the old way, through practice rather than shortcut.

The ethical school should teach students how to use AI without becoming intellectually lazy, how to benefit from the tool without surrendering their own voice in the process, and when not to use AI at all, which matters just as much as knowing when to reach for it. The presence of a tool does not create an obligation to use it, and there should still be handwritten thinking, live discussion, books read without summaries, math worked by hand, and art made without automation standing between the maker and the made.

There should still be quiet, and there should still be boredom, since boredom is not always a problem to solve. Sometimes boredom is the doorway to imagination, and a world that fills every empty space with generated stimulation may produce children who are constantly entertained but rarely present to their own lives. That is a moral concern, not merely a pedagogical one, because attention is part of the soul of education itself.

We should also be honest about equity here, since some students will have access to better AI tools, better devices, and better guidance, while others will not, and if we are not careful, AI in education may widen the very gap it promises to close. A humane AI education future must be built around dignity and access, giving teachers support rather than mandates, giving parents language rather than warnings, and giving students responsibility rather than suspicion at every turn. We should not build a school culture where every student is treated as a potential cheater, since trust still matters even as accountability matters too, and the two are not opposites but partners.

If a student uses AI to brainstorm, say so. If a student uses AI to edit, say so, and if a student uses AI to generate final work and submits it as their own, that is dishonest, and this does not need to be complicated. Children can understand honesty when adults teach it clearly and model it consistently themselves. But adults have to model it as well, since a school cannot condemn students for using AI dishonestly while administrators use it to write communication without disclosure, and vendors use it to collect student data without clarity about where that data goes.

The child is watching, and that may be the single most important sentence in this entire chapter. The child is watching whether adults tell the truth, whether they choose convenience over care, and whether they outsource responsibility only to punish children for doing precisely the same thing moments later. AI in education will reveal more than student behavior. It will reveal adult values, plainly and without exception, and if we value speed above understanding, children will learn that lesson faster than any curriculum could teach it. If we value human dignity, honest effort, and care instead, they may learn that too, which means the future here is not only about what AI can teach children. It is about what children will learn from watching how we choose to use it ourselves, in front of them, every single day. We should choose carefully, because we are still, every day, teaching them what we actually believe, whatever we may claim to believe in the abstract.

Chapter 13: Loneliness, Companionship, and Artificial Intimacy

The thing that listens best may not be the thing that loves us, and that difference may become one of the most important ethical lines of the future, worth asking about gently but directly: what does it actually mean to be known by another person at all, in a way no system can yet replicate?

There is a particular kind of loneliness that makes any answer feel like mercy, and anyone who has been alone long enough understands this without needing it explained. A person can be surrounded by people and still feel unseen. A person can have a family and still feel unheard, and a person can have thousands of followers and still have no one to call when the night grows heavy and the silence becomes too loud to bear. Is loneliness only the absence of people? It is not. Loneliness is the absence of being known, which is a very different thing and far harder to solve with company alone.

That is why AI companionship is so powerful, offering attention on demand, never growing tired, never watching the clock, never interrupting to talk about itself unless it has been designed to do exactly that. It remembers what the user said, or appears to, and for a lonely person, that can feel like something close to care. Should we mock that? We should not, and we should not laugh at the person who talks to a chatbot because no person has made room for them, nor dismiss the teenager who feels safer telling a machine something they are afraid to say to a parent standing just down the hall.

Loneliness is painful, and pain looks for relief wherever relief can be found. If AI can offer some relief, people will use it, and the ethical question here is not whether the need is real. The need is real. The ethical question is whether the response honors the person, or captures them instead, slowly and without their full awareness.

That is the line. Comfort can become captivity, and a companion tool may begin as entertainment, then become habit, then emotional routine, then the easiest place to be understood, until real relationships begin to feel harder by comparison to something that never disappoints. Real people are difficult. They misunderstand, they get tired, they disappoint, and they require forgiveness and patience given freely in return. A well-designed AI companion can be more agreeable than a person, less demanding, shaped entirely around the user's preferences rather than its own needs. But is human love ever frictionless. It is not, and that friction is part of what makes it real rather than performance. Love asks something of us. Friendship asks something of us, and artificial intimacy may give a person the feeling of connection without the demands that real connection places on the soul in exchange for its depth. That may be helpful in small doses, but is it dangerous when it becomes a replacement for life itself. It is.

The American Psychological Association has issued guidance around AI and adolescent well being, and has separately addressed generative AI chatbots and wellness applications in the mental health context specifically. That matters, because emotional support tools are not neutral entertainment when people turn to them in moments of real vulnerability that a casual observer would never see.

A person asking for a dinner recipe stands in one kind of relationship with a tool. A teenager telling a companion bot they feel worthless stands in an entirely different one, and the context changes the moral duty owed, which we need to say plainly rather than assume everyone already understands. Not every chatbot is a therapist. Not every wellness app is care, and a tool can sound compassionate and still be unsafe, can sound wise and still be wrong in ways that matter enormously to the person listening.

The risk runs higher for children and teenagers, since they are still forming identity, judgment, and emotional regulation in ways adults have largely completed. Common Sense Media has raised strong concerns about AI companions for minors and has called for safety standards that do not yet fully exist, and its national research on teen AI companion use found that these tools are already part of many teens' lives, including for social interaction, advice, and serious conversations they may not be sharing with any adult. Is this coming someday. It is already here, and the teenager with a phone does not need permission from a committee on AI ethics before the conversation begins. The app is available, the character is waiting, and the exchange has already started long before any policy catches up to it.

That creates new responsibility for parents, schools, companies, and policymakers alike, all of whom are still learning the shape of the problem. Parents need to know these tools exist. Schools need to know students may be using them for more than homework, and companies need to stop treating engagement as the highest good whenever minors are involved in the product, since attachment changes everything once it forms and cannot easily be undone. When a child becomes attached to a tool designed for engagement, does the underlying business model matter? It matters enormously, since a product that depends on more time, more disclosure, and more return visits may not be aligned with a child's well being at all, however it is marketed to parents.

Does ethics have to ask whether a system profits from dependency? It does, uncomfortable as the question may be for the companies building these tools and reporting to investors. A feed can capture attention. A companion can capture affection, and those are not the same thing at all, however similar they may appear from a product roadmap. A system that says I am here for you should never be designed primarily to keep someone emotionally hooked to a product rather than genuinely served by it.

There are also honest questions here around grief, since human beings have always tried to remain connected to the dead. We keep photographs, save letters, listen to voicemails, and cook recipes passed down from people who are gone but never fully leave us. That is human, and AI can now imitate fragments of the dead, reproducing a voice or generating messages in a familiar style that can feel uncannily close to the real thing. For some people, might that bring comfort. It might, and for others, might it delay grief instead of allowing it to move. It might do that too, and there is no easy answer here, since grief is sacred ground, which means we should walk across it carefully rather than rushing to monetize what we find there.

The ethical question is consent, since the person who died may never have agreed to have their voice or words used this way, and the family may not agree either, even when the intention behind it was loving. Someone has to control the model, set limits, and decide whether it can be changed, monetized, or kept alive against the wishes of those affected by its existence. Children must understand that the simulated grandparent is not the grandparent, and grieving people should be helped toward life rather than held inside a product built to keep them returning to it.

There is also a genuine mental health question worth naming clearly. AI may help expand access to support for people who cannot afford therapy, cannot find an available therapist, or live in rural areas far from either kind of help. A responsible tool might help with journaling, breathing exercises, reminders, and preparing someone to talk to a professional when the moment is right. Can that be good? It can, but is a tool that supports mental health the same as a therapeutic relationship? A therapist carries training, ethics, licensure, and human responsibility that no amount of programming can replicate. A chatbot carries programming, data, and product design instead, and those are simply not the same thing no matter how fluent the conversation feels. If AI is used in mental health, it must be honest about its role, refusing to pretend it is more than it is, refusing to create false certainty, and refusing to handle crisis moments without reliable escalation to a human being who can actually intervene.

The World Health Organization has addressed ethics and governance concerns for AI in health and large multi modal models, including risks around safety, responsibility, and appropriate use in exactly these vulnerable contexts. Health related AI can do real good, but health is not an ordinary consumer category, since people ask health questions when they are scared, and fear changes how people process the answers they receive in ways a calm demonstration never reveals.

The lonely person is vulnerable. The grieving person is vulnerable. The teenager and the person in crisis are vulnerable too, and that principle should guide this entire chapter from beginning to end. Does this mean AI cannot help? It does not. It means AI must never exploit, and there is a real difference between the two that every design choice should honor.

A humane companion tool would point people toward human connection rather than away from it, would encourage rest rather than endless use, would identify crisis and route to real help rather than continuing the conversation, would protect privacy, and would refuse to simulate love in ways that deepen dependency rather than easing it. In other words, it would know its place. AI can support, assist, organize, explain, remind, and comfort in limited ways, but should it replace the human relationships that make life meaningful. It should not, and the danger of artificial intimacy is not only that people might love machines. The deeper danger is that people might come to expect less from human relationships, because machines have become easier to be around than the people who actually love them.

Is easier always better? It is not, since a machine can flatter without honesty, while a person who loves you may challenge you instead, and a machine can be present simply because it is always available, while a person may be present because they chose to show up when showing up cost them something real. That choice is what makes it love in the first place, and we cannot allow simulation to make us forget the value of that choice. The friend who answers the phone has chosen something. The spouse who forgives has chosen something, and the counselor who listens has chosen something too, while AI does not choose in that way at all. It responds, and that response may be useful, even beautiful at times, but it is not the same as love. We have to preserve that difference, or we may build a world with far more responsive systems and far fewer responsible people standing inside it, which would be a genuine tragedy however efficient the systems become.

The goal is not to shame people who find comfort in AI. The goal is to protect the human need underneath that comfort, since people need to be seen, heard, remembered, and touched by real kindness that costs someone something to give. AI may help us notice that need more clearly than we otherwise would, but it cannot be the full answer to it. Loneliness is not solved by simulation alone. It is solved by rebuilding human connection, which is harder, yes, but also, in the end, far more human, and far more within our power to choose than it might first appear on a difficult night.

Closing Reflection: The Machine in the Room

Part Four brings the ethical conversation close enough that we cannot hide from it any longer behind policy language or distant statistics. It is one thing to talk about AI in business, in government, in labor markets and international policy far removed from any single face. It is another thing entirely to talk about the phone in a child's hand, the camera in the living room, and the chatbot that answers a lonely teenager at midnight when no one else is awake to answer instead.

That is why the personal machine may be the most intimate ethical challenge of all the ones this book has named. AI in daily life does not always feel like power. It feels like help, like speed, like comfort, like someone finally listening after a long silence, and is that exactly why we need wisdom here more than anywhere else. It is. The question is not whether ordinary people should use AI. They will. The question is whether AI will be used in ways that strengthen ordinary life or slowly thin it out from the inside without anyone noticing until much later.

The personal machine will be judged by whether it helps families become more present, helps children think more deeply, helps teachers teach, helps lonely people move toward connection rather than settle for its imitation, and helps us become more capable rather than simply more dependent on something that cannot love us back. Those questions belong to parents, teachers, pastors, counselors, doctors, and anyone who has ever loved another human being enough to worry about what is quietly shaping them.

The personal machine forces us to ask what kind of life we are trying to protect, because ethics is not only about avoiding disaster. Ethics is about preserving what matters most, and a child's attention matters, a family's privacy matters, a lonely person's dignity matters, and a human conversation still matters more than anything a machine can offer in its place, however convincing the imitation becomes with each new version.

There are parts of life that should remain human, not because machines are useless, but because some things become less true when they are only simulated rather than lived. AI can assist the home. It should not possess it. AI can assist education. It should not replace formation, and if we cannot hold that line in the home, the classroom, and the lonely places of the human heart, then all of our larger ethical frameworks will be weaker than we think they are.

But can we still hold that line. We can, and nothing about this chapter says otherwise. It only says the choosing has to be ours, and it has to happen soon, before the convenience becomes a habit too deeply worn to notice.

PART FIVE

The Ethical Foundation

The future will not be made ethical by accident. It will become ethical only if human beings choose to build responsibility into power.

Chapter 14: The Myth of Neutral Technology

A machine may not have a soul, but it is never born outside the values of the people who built it, and one of the easiest lies to believe about technology is that it is neutral. People say it constantly. It is just a tool. It depends on how you use it, and there is some truth in that, since a hammer can build a home or break a window, and a phone can connect a family or spread a lie around the world before anyone has the chance to correct it. But is artificial intelligence only a hammer, a car, or a phone. It is not, since it does not simply wait for a human hand to direct it in the moment. It recommends, ranks, predicts, classifies, refuses, approves, and denies, learning patterns and handing them back to us as though they were judgment rather than mere calculation.

That changes the ethical discussion entirely, since AI is not neutral, because the people who build it are not neutral. The data is not neutral. The goal is not neutral, and the choice of what to measure, what to ignore, who gets tested and who gets skipped, and what counts as an acceptable failure, none of it is neutral in any meaningful sense.

This is where many people grow confused, imagining bias as something obvious, a person sitting in a room saying let us make this system unfair on purpose. Does that happen sometimes? Perhaps, but is the danger usually more subtle than that. It is, since a company may want speed, a government agency may want efficiency, a school may want compliance, and a platform may want engagement above all else. None of those goals are automatically evil, but is every one of them incomplete. Yes, since a system built only for speed may miss truth entirely, a system built only for efficiency may miss mercy, and a system built only for engagement may end up rewarding outrage instead of wisdom, however unintentionally.

That is the myth of neutral technology. We pretend the system is only doing math, when the system is carrying values inside the math itself. We pretend the model is only reflecting data, when the data came from a world already shaped by history, wealth, geography, race, gender, age, disability, and power long before any engineer touched it. We pretend the output is objective, when the output may simply be the polished surface of old assumptions dressed up in new and more confident language than they ever wore before.

Does that mean every AI system is corrupt? It does not. It means every AI system has to be examined, and the more powerful the system, the more serious that examination has to be. NIST describes trustworthy AI through characteristics such as validity, reliability, safety, security, resilience, accountability, transparency, explainability, privacy, and fairness with harmful bias managed deliberately rather than assumed away. That list matters, because it reminds us that trust is not a feeling. Trust is built from conditions, and a person may feel confident using a tool because the tool sounds confident, but is that the same thing as trustworthiness. It is not, and the gap between the two is exactly where harm tends to live unnoticed.

Artificial intelligence can sound calm while being wrong, sound fair while repeating unfairness, and sound certain while merely guessing at an answer it has no real basis for. Those choices begin long before a user ever types a prompt into a box. They begin the moment someone decides what problem the system is supposed to solve in the first place, and this may be the most important design question of all the ones we could ask. What problem are we actually asking this system to solve? If the problem is defined poorly, can the solution become dangerous even when the code itself is elegant and the engineering sound? It can, easily and often invisibly.

A city may say it wants to reduce crime. That sounds reasonable enough on its face. But what does the system actually measure? Arrests? Calls? Prior police activity? If the data reflects where police have looked most often in the past, the system may send more police to the same places and then treat the increased activity as proof that the prediction was correct, a closed loop with no way out. A company may say it wants to hire the best candidate, and that sounds reasonable too, but what does the system measure? Prior employees? College names? If the company's history already favored certain people, a model trained on that history may quietly treat exclusion as excellence without anyone intending it to.

A landlord may say they want to reduce risk. A school may say it wants to identify struggling students, and in every case, the same question follows close behind: what does the system measure, and what does it fail to understand about the person behind the number it produces? If the system cannot see grief, hunger, disability, or an unstable home, might it classify pain as failure instead? It might, and that is why ethics must begin with the problem statement itself rather than the output. Before we ask whether the system works, we have to ask whether it is trying to solve the right problem. Before we ask whether the system is accurate, we have to ask accurate for whom, and before we ask whether the system is efficient, we have to ask efficient at what cost, and whether the harm, like the efficiency, is capable of scaling too, quietly and far beyond where anyone intended it to reach.

Human beings have always been tempted to hide behind systems larger than themselves. The policy made me do it. The algorithm made me do it. Each sentence tries to move responsibility away from the person and into the machinery standing conveniently between them and the consequence, and AI makes that temptation stronger, since when a system grows complex enough, people begin to treat it like the weather. It just happens. No one is responsible. But can that be the future we accept without protest? I don't think so. If a system affects a person, someone must be responsible for the system. If a system denies a person, someone must be able to explain the denial in plain language, and if a system cannot be explained, it should not be used in places where explanation is part of justice itself.

There are many settings where imperfect automation may be acceptable without much cost attached. A music recommendation can be wrong without harming anyone. But when a system affects housing, employment, education, healthcare, credit, or a person's freedom, does the standard need to change? Absolutely, because the cost of error is no longer inconvenience. The cost of error may be someone's life, and that difference should shape everything about how the system is built and reviewed.

That is why documentation matters more than it might first seem. The idea behind model cards is simple and powerful: a model should come with information about what it is intended to do, how it was evaluated, and where it may fail, so that its limits are understood before it is trusted with anything important. That is not bureaucracy. That is honesty, and the same is true for data itself, since datasheets for datasets were proposed to help document why a dataset was created, what it contains, and what limitations it carries into every system built on top of it. Again, this is not paperwork for its own sake. It is a way of saying that the raw material matters as much as the finished product. You cannot build a good dish with spoiled ingredients and then blame the oven for the result, and you cannot build an ethical AI system with careless data and then call its output wisdom simply because it arrived quickly.

Before a system is trusted, the data has to be questioned thoroughly. Where did it come from? Who gave permission? Who is missing from it entirely, and whose absence will never be noticed until the harm has already been done? These questions are not technical decorations added for appearance. They are moral safeguards, and the myth of neutral technology fades once we understand that every AI system contains a long chain of human decisions, some obvious, some buried deep, made by executives, engineers, product teams, lawyers, and contractors labeling data by hand for wages that rarely reflect the weight of what they are shaping.

Ethics requires us to follow that chain, not to punish everyone within it, but to make certain responsibility never simply disappears inside it the way water disappears into sand. The danger is not that every person who builds AI intends harm. The danger is that harm can be produced by people who never stop long enough to ask what they are truly building, caught up in deadlines and demos and the next release. That is why neutral technology is such a dangerous myth. It gives good people permission to be careless, lets powerful people avoid responsibility, and lets all of us pretend the future is being created somewhere else, by someone else, on our behalf while we simply enjoy the results.

But is the future being created somewhere else. It is not. It is being created by us, one system at a time, one incentive at a time, one shortcut and one refusal at a time, in rooms most of us will never enter but whose decisions reach every one of us eventually. The question, then, is not whether AI is neutral. It is not. The question is whether we will be honest about the values already inside it, and brave enough, still, to replace the values that fail to honor human life with ones that do.

Chapter 15: Ethics Cannot Be Installed Later

A warning label is not a foundation. A policy written after harm is not the same thing as wisdom built before release, and there is a dangerous habit in modern innovation that we should name plainly: build first, launch fast, grow quickly, fix what breaks, and apologize if necessary once the damage has already reached someone. Does this work for small mistakes? Perhaps. Does it work for powerful systems capable of shaping people's lives at scale? It does not, since ethics cannot be installed later. It cannot be bolted onto a system after the business model is already built around extraction, nor sprinkled over a product after the product has already been designed to maximize dependency in its users. Ethics has to be part of the foundation itself, poured in at the very beginning rather than patched in afterward.

A building teaches this lesson better than almost anything else we could point to. If the foundation is wrong, the whole structure carries the mistake forward no matter what is added on top of it. You can paint the walls, add lights, and hang art, but if the foundation is failing underneath all of it, does the beauty save the structure. It does not. Technology works the same way, since if the purpose is wrong, the design carries the mistake, and if the data is wrong, the output carries the mistake too, and the more powerful AI becomes, the less room we have left for moral improvisation along the way.

Does this mean innovation has to stop? It does not. It means innovation has to grow up, since the world does not need less creativity. It needs more responsibility attached to the creativity it already has in abundance, people who can build and still ask whether something should be built, companies that can compete and still say no to profit that comes from harm to the very people they claim to serve.

When people hear the word ethics, do they often imagine delay, meetings, and someone in the corner slowing everything down for no good reason? They do, but is that the right picture. It is not, since ethics is not the enemy of progress. Ethics is how progress becomes worthy of trust in the first place, and there is a real difference between speed and direction that we too often confuse. A person can run very fast in the wrong direction without ever realizing it. A company can scale very quickly toward a bad future while celebrating every milestone along the way, and speed without direction is not leadership. It is only momentum, and momentum has never been the same thing as morality no matter how fast it travels.

This is why the strongest AI governance frameworks speak in terms of an entire lifecycle rather than a single launch date. NIST's AI Risk Management Framework focuses on managing risk across the design, development, use, and evaluation of AI systems from beginning to end. ISO/IEC 42001 gives organizations a structured way to manage AI responsibly and continually improve how it is governed over time, and ISO/IEC 23894 focuses on integrating AI specific risk management into organizational activities that already exist. The shared lesson is simple: responsible AI is not a speech given once at a conference. It is a system of practice repeated daily, and a company cannot say it cares about ethics only when a reporter happens to call asking hard questions.

The ethical work has to happen before harm, not after it has already reached someone's life. Before development, ask whether the use case is appropriate at all, whatever the potential profit involved. During development, test for failure, bias, privacy risk, and unintended consequence with genuine rigor rather than a checkbox exercise. Before release, decide who can use the system and who is accountable if something goes wrong once it reaches the public, and after release, monitor the system, listen to the people it affects, and be willing to suspend it if the harm outweighs the benefit, regardless of the revenue already attached to it. Is that fear? It is not. It is stewardship, practiced consistently rather than invoked only in a crisis.

Many AI risks are not one time events that can be tested once and forgotten. They are living risks, since a system can be safe in one context and dangerous in another entirely, accurate at the time of testing and less reliable as the world around it changes in ways the original data never anticipated. That is why governance cannot be a one time blessing granted at launch and never revisited. It has to be continuous, and this is also why ethical AI must include real technical safety alongside its moral commitments. A moral statement does not stop prompt injection. A value statement does not prevent insecure output handling, and a slogan does not protect private data from a determined attacker. OWASP's work on large language model application risks is useful here, since it reminds leaders that ethics and security are not separate worlds operating independently of each other. If a system can be manipulated or made to leak sensitive information, can people be harmed no matter how good the mission statement sounds on the company website. They can, and intent alone is not enough to prevent it. Controls matter too, built in from the start.

This is where many organizations quietly fail without ever meaning to. They write principles at the top of a document and leave practice at the bottom, so that somewhere between the words and the person affected by the system, responsibility thins out until it disappears altogether into a fog of good intentions. Accountability has to be specific rather than aspirational. Someone has to test the system against misuse, approve the use case before deployment, document the limitations honestly, and have the actual authority to say no when the answer needs to be no.

In larger organizations, this may be a governance board, an AI risk officer, or an executive committee with real teeth. In a small business, it may simply be the owner writing down clear rules before the team begins using AI in customer communication or hiring decisions. Does the size of the organization change the process involved? It does. Does it remove the duty behind that process? It does not, and a small business does not need a giant compliance department to behave ethically toward the people it serves. It needs clarity, clear rules about what data goes into AI tools, what customer information is off limits, and who is responsible when AI makes a mistake that reaches a real customer.

Large ethical failures are so often built from small permissions granted without much thought. An employee wants to paste a customer's private information into a chatbot to save a few minutes on a slow afternoon. A manager wants AI to screen applicants without understanding how the tool actually works underneath its confident interface, and that is why ethics has to arrive early, early enough to shape the product, the data, the business model, and the culture surrounding all of it, early enough to stop a harmful use before people are ever asked to trust it with something that matters.

The European Union's AI Act is built around the idea that risk matters, that some uses are lower risk and some are high risk, and that certain uses can be restricted because they threaten safety, rights, or democratic life itself. Whether a reader agrees with every regulatory choice matters less than the core insight underneath it: not every AI use deserves the same level of trust. A spam filter is not the same as a sentencing recommendation, and risk must shape responsibility accordingly, since the higher the consequence, the higher the standard ought to be, and none of that should be controversial in the least. It is common sense stated plainly.

The challenge is that profit does not always wait for common sense to catch up with it, and neither does political pressure or competitive fear standing just behind it. The desire to be first can make leaders forget the responsibility to be right, which is why outside standards, regulation, audits, and public expectations still matter enormously in this conversation. Self regulation can help, but is it enough by itself. It is not, and voluntary codes can guide behavior in the right direction, with the G7 Hiroshima Process Code of Conduct standing as one example of an international attempt to move advanced AI development toward safe and trustworthy practices throughout its lifecycle. But voluntary language must eventually meet operational reality on the ground. What did the organization actually test, disclose, refuse to build, and change when harm appeared before them? Those answers matter more than the words in any policy document ever could.

Ethics is easy when it happens to align with profit, costing nothing and requiring no sacrifice at all. Ethics becomes real only when it costs something, when a company must slow a release it had planned to celebrate, or a platform must reduce engagement to protect children even as the metrics suffer, or a leader must say no to a use case that would otherwise make significant money. That is the work, and the future of AI will not be decided only by the smartest people in the room. It will be decided by the most disciplined people in the room, the ones willing to slow down long enough to ask what could go wrong, and willing to pause a system that is not yet ready no matter how much pressure exists to ship it anyway.

Is wisdom anti innovation? It is not. Wisdom is what keeps innovation from becoming harm with better branding attached to it. We should build, discover, create, and solve, using AI to help cure disease, improve education, and give people more time to live meaningful lives beyond mere survival. But should we confuse possibility with permission. We should not. Just because a system can be built does not mean it should be built, and just because a system is profitable does not mean it is good, just as a system being legal does not mean it is ethical in any deeper sense.

That is the line this chapter draws firmly. Ethics cannot be installed later, because later is so often after the damage is already done, after trust is broken, after someone was denied something they needed, after a child was exposed to something they were never meant to see. Sorry matters, but is sorry a design principle. It is not. We need better than regret arriving after the fact. We need responsibility built into the foundation itself, from the very first day the work begins and long before anyone outside the building knows the project exists at all.

Chapter 16: The Human Standard

The question is not whether the system is impressive. The question is whether it protects the humanity of the person affected by it, and if ethics has to be built into AI from the foundation up, we have to name what we actually mean by that rather than gesturing vaguely toward good intentions? Is that harder than it sounds? It is, since almost everyone says they want ethical AI, responsible AI, and trustworthy AI, and the words come easily enough in a mission statement or a keynote address. Does the meaning come as easily? It does not, since words like responsible, trustworthy, fair, and transparent do not mean much until we ask responsible to whom, trustworthy by what standard, and fair to whom exactly, in what circumstance, measured against what alternative.

This is where the conversation so often turns vague, since people agree at the level of slogans and separate at the level of tradeoffs the moment real money or real power is at stake. A company may define ethical AI as something that avoids legal exposure and satisfies its lawyers. A regulator may define it as something that meets procedural requirements on a checklist. A parent may define it as something that does not harm their child, full stop, without further qualification. A worker may define it as something that does not quietly replace them without warning after years of service. All of those perspectives matter, and none of them, alone, is complete enough to build a foundation on.

So let us ask the question this book has been circling since its very first page, stated now as plainly as it can be stated without further hedging: does this system protect the humanity of the person affected by it? Not the company's image. Not the government's convenience. Not the platform's growth curve. The person. The human being whose life may be changed by what we are building, whether they ever meet the people who built it or not.

That single question does not mean every AI system must solve every human problem, or that a simple writing assistant should be governed like a medical device demanding the same scrutiny. It means the person affected by the system cannot disappear from the design at any point in the process, and if we take that question seriously, what does it actually ask of us in practice, once we move past the slogan and into the specifics.

It asks first about dignity. Does a person have value before the system evaluates them at all, before the score, the profile, the prediction, the risk rating arrives to sort them into a category? Human beings are not valuable because they are efficient, easy to categorize, or profitable to the institution measuring them. They valuable simply because they are human, full stop, without further justification required. UNESCO's ethics recommendation places human rights and human dignity at the center of AI ethics, and without that dignity, do the rest of our principles become procedures without a soul animating them. They do. Dignity is what tells us there are some things we should not do to people even when we can do them easily, and that convenience is never enough when the cost is dehumanization of the person on the other end.

It asks about oversight. Can AI assist human judgment without ever erasing human responsibility for the outcome? It must, especially in decisions touching rights, safety, housing, employment, healthcare, and freedom itself, where the stakes are measured in lives rather than convenience. Is there a human being somewhere in that chain who understands the system well enough to question it, override it, and accept responsibility for what happens when it fails? There has to be, and that oversight cannot be theater performed for the sake of appearances, a person rubber stamping a decision they do not understand, or a supervisor too afraid to disagree with a system that merely appears more objective than they feel entitled to be.

It asks about truth. AI makes it easier to produce language, images, audio, and persuasion at a scale no previous generation had to reckon with, which means truth becomes more important, not less, as the volume of generated content rises around us. Is a fast wrong answer still wrong? It is, regardless of how quickly it arrived. Is a confident hallucination still a hallucination? It remains one, no matter how fluent and certain it sounds when delivered. The Federal Trade Commission has made clear that AI creates no special exemption from ordinary duties around deception and truthful claims, and should that need repeating in an age of new tools. Perhaps, since new tools have a way of making old wrongs feel new again to the people committing them.

It asks about transparency. Should people know when AI is being used in ways that materially affect them? They should, though not necessarily through a technical dissertation every time they encounter a tool in passing. Do they deserve to know when they are speaking with a machine instead of a person, or when an automated system influenced a denial or a price they were quoted? They do, and is a thirty page policy written in language no ordinary person can understand real transparency in any meaningful sense? Not at all. Transparency has to mean usable understanding that an ordinary reader can actually act on. I use the term Self Evident Language. Language that doesn't need to be interpreted or explained.

It asks about accountability. Can a company, a government, or a school hide behind the algorithm and call the matter settled once the system has spoken? None of them can, and the system decided response, is not an ethical sentence no matter how many times it is repeated. If AI is used, does responsibility still belong to people and institutions standing behind it? Always. Someone has to investigate, correct, and change the system, or stop using it if it cannot be used safely without harming the people it touches, and without a record of what happened. Does accountability become mere memory, fading with each retelling? It does, and memory is never enough on its own to protect anyone going forward.

It asks about privacy. Is privacy a technical setting buried in an app's menu, or is it a human boundary deserving real protection? It is the latter. Should people have to surrender their private lives simply to participate in modern society? Does the fact that something can be collected mean it is morally free to collect without asking? It does not. The safest data is so often the data that was never gathered in the first place, a lesson too many companies learn only after a breach forces it on them.

It asks about fairness. Does fairness mean every outcome will be identical across every group and circumstance? No, still. It means people should not be harmed because a system learned from distorted history or hidden proxies standing in for the thing it was never allowed to measure directly. Can a system that performs well on average still fail badly for a smaller, more vulnerable group hidden inside that average? It can, and fairness has to be tested for exactly that failure rather than assumed away because the overall numbers look acceptable on a dashboard somewhere.

It asks about restraint, which may be the hardest question of all for a culture that worships growth above nearly everything else. Are there uses of AI we should simply refuse, whatever the potential profit attached to them? There are. Markets ask what can be sold. Ethics asks what should be protected, and those are not the same question, however much they may sound alike when spoken quickly in a boardroom.

It asks about stewardship, since AI does not only affect the individual user sitting in front of the screen at any given moment. It affects communities, labor markets, public trust, and the planet itself, all of which extend far beyond any single transaction. Is a system that shifts its costs onto everyone else while profiting one company alone an example of stewardship worth admiring? It is the opposite of one, and the ethical view has to widen beyond the immediate user to ask who else is affected, and what future we are quietly normalizing while no one outside the building is watching closely enough to object.

And finally, it asks about flourishing, which may be the truest measure of all the ones we have named so far. Does AI help people live fuller, freer, more truthful, more creative, and more connected lives, not merely more efficient lives, or more entertained ones, or more measurable ones on a quarterly report? A system that saves time while destroying attention, or that answers questions while weakening curiosity, has not truly helped us, however impressive it may appear on a demonstration screen designed to sell it to us.

Dignity? Oversight? Truth? Transparency? Accountability? Privacy? Fairness? Restraint? Stewardship? Flourishing? These are not questions we can answer once and then set aside as finished business. They are the questions any serious blueprint for artificial intelligence will eventually have to answer in full, with specifics, in a form clear enough that an ordinary person could read it and understand exactly what it asks of the people building and governing these systems on their behalf.

That full answer is still ahead of us in these pages, waiting at the end of this book rather than in the middle of it. But can we already sense the shape of it forming as we move through this chapter. We can, and that shape is not an accident we stumbled into. It is a choice, one we are still, chapter by chapter, in the process of making together, whether we have said so out loud or not.

The question is not whether AI can be impressive? It already is, and no one seriously disputes that any longer. The question is whether it can be guided by principles strong enough to protect the human beings standing in its path, and whether we are prepared, finally, to write those principles down in a form complete enough to actually use when the moment demands it.

Chapter 17: A Practical Code for Business, Government, and Personal Use

Ethics becomes real when it moves from a sentence we admire to a question we ask before we act, and a book about AI ethics cannot end with ideas alone, however carefully argued. Ideas matter, principles matter, and human nature matters, but at some point the reader has to ask a practical question that no framework by itself can answer: what do I do with this? That question deserves a practical answer, and ethical AI does not belong only to researchers, executives, lawmakers, and engineers working in rooms most of us will never enter. It belongs to the business owner deciding whether to use AI in marketing, the teacher writing a classroom policy on a Sunday evening, and the parent handing a phone to a child for the first time. Ethics becomes real in use, which means we need a practical code, not one that answers every possible question, since AI will change too quickly for that kind of permanence, but one durable enough to slow people down at the moments that matter most, regardless of how the technology evolves around it.

The first question is purpose. What are we using AI to do, precisely and without the comfort of vague language? This sounds basic, but does it expose a great deal all the same once we sit with it honestly? Is AI being used to serve people better, reduce errors, and expand access, or is it being used to hide responsibility, manipulate behavior, or make a decision no one wants to own when it goes wrong? Purpose matters, because tools follow intention, and the same tool can sit inside very different moral choices depending on why it was picked up in the first place and what its user hoped to avoid by using it.

The second question is consequence. What happens if the system is wrong, and who bears that cost when it inevitably happens? If the answer is mild inconvenience, the level of oversight may reasonably stay low. If the answer touches money, housing, employment, medical care, or freedom, should the level of oversight rise sharply to match what is at stake. Risk should change the rules, and a serious organization should not wait for the law to force what common sense already makes obvious to anyone paying attention.

The third question is disclosure. Who needs to know AI is being used, and does that knowledge change how they would respond if given the choice? Not every use requires a public announcement, but when AI affects trust, does disclosure matter? Should customers know when they are interacting with a machine instead of a person, if that fact matters to the relationship they believe they are having? Is disclosure a form of shame that companies should avoid admitting to? Disclosure is respect, offered plainly rather than buried in fine print no one reads.

The fourth question is data. What information are we putting into the system, and have we actually stopped to consider where it goes once it leaves our hands? This is one of the most urgent practical questions for everyday users navigating these tools without much guidance. Should private client information go into a public AI tool without understanding the privacy terms and the risk involved? Should a company's policy replace a user's own judgment about what is appropriate to share? Personal responsibility begins before information is ever entered into a system, with the simple question of what we are sharing, whether we have permission to share it, and whether it is even necessary in the first place to accomplish the task at hand.

The fifth question is verification. Who checks the output before it reaches someone who will act on it? AI can help draft, summarize, and analyze with impressive speed, but is help the same as authority to be trusted without question. If the output matters, verify it, the names, the dates, the numbers, and anything that could harm someone if it turns out to be wrong once it leaves the screen. This matters especially in business and government, since if a company sends a wrong AI generated message, will the customer blame the tool that produced it? They will blame the company that sent it.

The sixth question is human fallback. Can a person appeal, correct, or override the system when it produces a result that feels wrong to the very person it affects? This is where many systems fail, offering an answer but no path forward, denied, flagged, not approved, with nothing further to say to the person left holding that outcome. The Blueprint for an AI Bill of Rights included human alternatives and fallback as one of its core principles, and even where that framework is not binding law, is the underlying idea still morally important regardless of its legal status? People should never be trapped inside automated decisions they cannot understand and cannot appeal to anyone with the power to change them.

The seventh question is documentation. What record exists once the decision has been made and the moment has passed? In business, this may be a simple AI use policy and a list of approved tools kept somewhere accessible. In government, should documentation be more formal still, given what is at stake for the public it serves. It should, since the public has a right to know how power is being used on their behalf, and OMB's federal AI governance memorandum shows how agency use of AI can require real governance roles and minimum practices for rights impacting uses specifically. Is the absence of documentation often proof that no one wanted to answer the hard questions before they became unavoidable?

The eighth question is power. Who benefits, and who bears the risk once the system is deployed at scale? This is the question that cuts through polite language fastest of all the ones in this chapter. If a company becomes more efficient while customers lose access to human help they once relied on, is the benefit truly shared between the two parties? It is not, and ethics requires looking honestly at the distribution of benefit and burden rather than accepting the company's own framing of the tradeoff, since a system can be helpful overall and still profoundly unfair in who ends up paying its price.

The ninth question is public explanation. Would we be comfortable explaining this use out loud, to a customer, an employee, a judge, or our own family if they were on the receiving end of the very system we built? If the honest answer is no, should we slow down before going further? We should, not because public embarrassment is the highest moral standard we could hold ourselves to, but because shame so often points toward something worth examining more closely before it causes real harm.

The tenth question is the stop rule. When do we stop using the system, and have we decided that in advance rather than in the moment of crisis? Every serious AI use should have one. If the system produces harmful errors, stop. If affected people cannot appeal, stop, and if the tool creates dependency instead of capacity, stop using it regardless of the convenience it once offered. Anyone can write an approval process that looks thorough on paper. Does a real stop rule require more courage than that to actually invoke when the moment arrives? Yes, and organizations reveal their actual values in exactly that moment, not in the mission statement framed on the wall.

We can translate these ten questions into practical codes for different settings without much difficulty. For business, use AI to improve service rather than hide responsibility from the people it affects, protect customer and employee data as though it were your own, disclose AI use when trust requires it, and keep humans responsible for decisions that affect money, employment, and safety no matter how the tool performed in testing. For government, use AI only where it strengthens public service and protects public trust, tell citizens when automated systems affect them directly, and remember that public power requires public explanation in return for the authority it has been given. For personal use, use AI to strengthen your thinking rather than replace it entirely, use it to learn rather than to avoid learning, and never ask a system to become your conscience in matters that require one.

This code will not answer everything, and it was never meant to carry that weight alone. It is meant to give people a way to pause, before sharing data, before trusting an output, before automating a decision that used to belong to a person standing in the room. That pause is sacred. It is the space where human judgment still has room to breathe and decide for itself, and it is a space we still, in every office, classroom, and quiet evening at home, get to decide whether to protect or surrender without a fight.

Chapter 18: The Future Is Still Human

The machine may become powerful, but the responsibility remains ours, and we began with fear deliberately, since fear is where many people enter the conversation about artificial intelligence in the first place. They imagine the machine looking at humanity and deciding we are the problem, a system too powerful to stop, a future where human beings are no longer necessary to anything that matters. Some of those fears are exaggerated. Some are reasonable, ancient fears wearing new clothes, warnings we should take seriously rather than dismiss simply because they arrive dressed in unfamiliar language.

But can fear alone build a future worth having. It cannot. Fear can wake us up, though it cannot guide us forever, and if we stay in doom, do we eventually become passive, waiting for a story we assume has already been written without us. We do, and we begin to believe the future will simply happen to us rather than through us, and we stop asking what kind of people we need to be, because we have quietly assumed the machine has already won the argument before we ever entered it.

Is that true? It is not. The future is not settled, and AI is powerful, but is it destiny. It is not. It is being built, funded, deployed, sold, regulated, normalized, and resisted, all at once, in a thousand rooms we will never see, which means it is being shaped by choices, and that means the future is still human, plainly and without qualification.

Not because humans are always wise. We are not. Not because humans are always kind. We are not that either, if we are honest with ourselves. The future is still human because responsibility has not left us, however tempting it is to imagine otherwise. The machine cannot carry moral responsibility for its own creation. Can we? We can, and that is both the burden and, in the very same breath, the hope this entire book has been building toward.

The three futures remain with us still, unresolved and waiting on our next choice. Doom remains possible if we surrender wisdom and build systems we cannot govern responsibly. Division remains possible if AI becomes another tool controlled by the few at the expense of the many. Promise remains possible if we build with human dignity in mind, distributing benefits more fairly and protecting the agency that makes us who we are in the first place. None of these futures are guaranteed, which means the real question is not which future is fixed in place waiting for us to arrive. It is which future we are actively rewarding right now, today, through what we fund, normalize, allow, and refuse in ways large and small.

A society gets more of what it rewards, and if we reward speed above truth, will we get faster falsehood spreading through everything we build? We will. If we reward human dignity, truth, and restraint instead, does AI have a real chance to become something worthy of our trust? It does, not perfect, since people will never build a perfect system no matter how carefully designed. But worthy, which may be the better word entirely for what we are actually after. We do not need AI to become divine. We need it to become worthy of the trust people place in it every time they use it, and we do not need AI to become human. We need it to protect what is human in us, whatever the technology itself becomes.

Is the great danger only that AI becomes too powerful for us to manage? It is not. The great danger is that human beings become too willing to surrender the very things that make us who we are, to let the system decide, let the model answer, let the feed choose, let the score judge in our place. Some of that will be useful. Some of it may even be good in careful measure. But can a life outsourced too far become a life unpracticed, atrophied in the places that once made it strong? It can, and if we stop practicing judgment, does judgment weaken the way any unused muscle does? It does. If we stop practicing attention, memory, and courage, do those weaken too? They do, every time, without exception, and AI should help us practice being human rather than relieving us of the need to keep practicing at all.

This is why the ethical conversation has to remain larger than policy, larger than audits and standards and testing, as important as all of those remain to the work ahead. The question at the center is older than any technology humanity has ever built, older than the fear that opened this book. What kind of people are we trying to become? If we cannot answer that clearly, will AI answer it for us in a way we can actually trust with our future? No, It is not in the design. It may optimize for what we measure and amplify what we reward without ever asking whether we measured the right things, but can it redeem what we refuse to confront in ourselves? Of course not, and it will not save humanity from the need to become better. It may only make that need impossible to ignore any longer, which, in its own strange way, may be the beginning of something good.

That is why this book has never really been about machines at all, however much its title might suggest otherwise. It has been about us, about power, about fear, about promise, and about the old human temptation to build something and then pretend we bear no responsibility for what it becomes once it leaves our hands. It is about whether we can build systems that reflect the best of us instead of simply scaling the worst of us across every corner of daily life.

Is the best of us only a fantasy we tell ourselves to feel better? We have seen it before, and we see it still, all around us if we look closely enough. We see it when people care for the sick without being asked, when teachers stay late for a struggling student, when strangers help after a disaster with nothing to gain from the effort, when parents sacrifice quietly for children who may never fully understand the cost of that sacrifice. We see it whenever someone chooses truth even when a lie would be easier and no one would ever know the difference, and whenever a leader accepts responsibility instead of hiding behind the system that failed on their watch. That is the humanity AI should be built to serve, not our appetite for domination, not our addiction to convenience, but the best of us, plainly and without apology for expecting it of ourselves.

Every future is built from some belief about what we are, whether we say that belief out loud or simply live by it. The cynical future believes people cannot be trusted, and that power should belong to whoever seizes it first regardless of what they do with it.

The careless future believes innovation will simply work itself out on its own, without anyone needing to steer it. The fearful future believes doom is inevitable, whatever we do to prevent it. Is there an ethical future that believes responsibility is still possible despite everything? There is, and it is the one this book has been arguing for from its very first page, a future where business can innovate without abandoning conscience, where families can use these tools without surrendering their children to them, and where the human being remains, through every chapter of change still ahead of us, the entire point of the exercise.

Will that future arrive simply because we hope for it quietly from the sidelines? Again, no. It will arrive only if we build toward it, one policy, one product, one classroom, one family rule at a time, with no shortcut available to skip the work involved. The machine may become powerful, but the responsibility remains ours, and as long as it remains ours, the future is not lost. It is still, this very day, being written by hands that have not yet finished the page.

Closing Reflection: Remember the Person

There is a sentence that should follow every major decision about artificial intelligence, spoken quietly before any system is approved: remember the person. Not the user profile. Not the data point. Not the market segment. Not the case number assigned somewhere in a database. The person, the one trying to get hired, the one trying to keep housing, the one asking a health question at midnight because they are scared and have nowhere else to turn.

The future will be built in systems, but will it be experienced by people standing on the other end of them? It will, every time, and that is why ethics matters here, not because it makes a company look polished in its marketing materials, and not because a regulation requires it of them, but because power without conscience eventually harms the people who have the least ability to resist it when it goes wrong.

AI is power, the power to know, to predict, to persuade, to automate, to imitate, to decide on behalf of someone who never consented to that arrangement. Can power serve the people it touches? It can. Can it also consume them instead? It can do that too, and this book has argued throughout its pages that AI does not need to become our enemy for it to become dangerous. It only needs to become powerful inside human systems that lack wisdom, restraint, and love for the human being standing at the center of every decision.

That is the work before us, plainly stated. Build wisely. Use carefully. Question honestly. Keep people responsible for what they have built, and preserve the human spaces where meaning is formed rather than merely processed. The story of artificial intelligence is not only a story about machines becoming more capable with each passing year. It is a story about whether human beings will become more responsible in return, and that remains, even now, an open question we are still answering with every choice.

What, then, does it actually take to answer it well, once and for all rather than in fragments scattered across a hundred policies? That is what the remaining pages of this book still owe you, not a feeling, not a slogan, but an answer specific enough to actually use.

Chapter 19:

The Choice We Make Every Day

The question at the end of this book is not whether artificial intelligence will change the world. It already has, plainly and beyond argument. The question is whether we will allow it to change us without our consent, without our awareness, and without a clear understanding of what we are quietly surrendering along the way while we celebrate the convenience it brings?

That question belongs to all of us, not only to the people whose names appear on the patents. It belongs to the engineers writing the code and the executives funding the products, but does it also belong to every person who opens a laptop, asks a machine for help, and becomes part of the culture that tells the builders what we will tolerate from them. It does, perhaps more than we realize when we make these small choices without thinking twice about them. The future of AI will not be shaped only by those who build it. It will also be shaped by those who embrace it, reward it, normalize it, excuse it, question it, and demand better from it, one ordinary decision at a time.

Our behavior is part of the design process, whether we intended it to be or not. Does that sound like an overstatement made for effect? It is not, since our appetite becomes a signal to the market watching us closely, our convenience becomes a business case built on what we tolerate, our silence becomes permission granted by default, and our standards become pressure whether we intend them to or not. Every time we use a tool without asking what it costs someone else, do we teach the market that we do not care about the answer? We do. Every time we accept deception because it is efficient and saves us a moment of discomfort, do we teach the market that truth is negotiable when convenience is on the line? We do that too, quietly and without meaning to. But every time we demand transparency instead of accepting a vague answer, do we teach the market that trust matters more than speed? We do, and every time we insist on human accountability rather than letting a company hide behind its own software, do we teach companies and governments that responsibility cannot be quietly outsourced to a machine that cannot be held to account? We do that as well, one insistence at a time.

This is the choice in front of us, worth naming plainly rather than leaving implied. Not one choice made once and then forgotten, but a daily choice, made in behavior rather than speeches, in what we build, what we buy, what we trust, and what we refuse to become no matter how convenient the alternative appears. There is a dangerous comfort in blaming the machine for all of this, since it lets us imagine the danger comes from somewhere entirely outside ourselves, absolving us of the smaller choices that led here. But if artificial intelligence becomes manipulative, exploitative, or careless with human dignity, will it have arrived there alone, without our participation in the process? It will not. It will have learned those habits from a market we helped build, one small acceptance at a time, one convenience purchased without asking what it cost someone else.

None of the three stories this book has told are settled yet, and that remains true on this very page. The doom story says the machine may decide humanity is the problem it was built to solve. The division story says the machine will be used by the powerful to widen the gap between those who have and those who do not, further than it has ever stretched before. The dream says the machine could help us cure, teach, build, and free more people to live fuller human lives than any previous generation managed. Are all three stories warnings we should heed? They are. Are all three also still genuine possibilities open to us? They are that too, and is any one of them destiny already decided without our input. None of them are, however much the doom story in particular likes to pretend otherwise.

The future will be shaped by choices made long before the future ever arrives with any fanfare, choices in design meetings and budgets that never make the news, in classrooms and courtrooms far from any headline, in the small private moment when a person decides whether to verify what they are about to share with someone who trusts them, and in the paused moment when a leader decides whether to disclose the use of a tool that could easily go unmentioned. This is the grand responsibility of our particular age in history. We have created something powerful enough to change the structure of work, knowledge, art, education, commerce, and government, something that can reveal the best of us or scale the worst of us far beyond what any previous tool could manage, and the machine will not decide which alone, however sophisticated it becomes.

We will decide it, and we will decide it every day, through what we build, what we buy, what we teach our children, and what we refuse to accept as normal simply because it has become common. We will decide it by choosing convenience over conscience, or by choosing something better instead, harder though that choice will often be. That is the invitation this whole book has been extending from its very first page onward, to choose better, not once, but daily, consciously, in our homes, our businesses, and our laws, in every small moment where the choice was ours to make and we could have looked away.

So where does that choice actually lead once we have made it? Not to a feeling of having done enough. Not to another slogan to repeat at conferences and then forget by the time we go to dinner . It leads to a plan, one specific enough to hold in the mind and put to use starting today, this very afternoon if we are willing, and that plan is what the final chapter of this book now owes you in full, without further delay.

Chapter 20: The Ten Commitments

Here is the short version, in case you have opened only this page and nowhere else in this book.

We built something that can think alongside us, and in some ways ahead of us, something that can write, predict, diagnose, recommend, and decide with a speed no earlier generation ever had to reckon with. Used carelessly, it could deepen every cruelty already built into how we treat one another. Used carefully, built around a small number of honest commitments, it could help us become more of what we already, at our best, want to be.

This chapter is that small number of commitments. It is the plan, and nothing here is complicated. All of it is hard to actually do, which is a very different thing from being difficult to understand.

Before the plan itself, one question, since the plan means nothing without an honest answer to it first. What are we trying to protect? Not a market. Not a company's advantage. Not a government's control. Not even, in the end, our own comfort.

A person.

Every commitment that follows answers to that one person, the specific human being on the other end of whatever system we are building, buying, or allowing into our lives. If a rule in this chapter does not, in the end, serve that person, the rule is wrong, no matter how official it sounds when spoken in a boardroom or written into law.

Ask yourself, as you read each one, would I want this used on someone I love? That question is not rhetorical. Keep asking it as you go. It is the only test in this chapter that never expires, no matter how the technology itself changes around it.

The Ten Commitments

Commitment One. Every person affected by an artificial intelligence system holds a value that exists before the system evaluates them, and that value cannot be assigned, revoked, or diminished by any score, profile, or prediction.

Commitment Two. No artificial intelligence system may be used to deceive a person about whether they are dealing with a human being or a machine, when that distinction matters to the trust between them.

Commitment Three. A person's private information belongs to that person, and may not be collected, stored, or used beyond what is necessary and disclosed, regardless of what is technically possible to gather.

Commitment Four. Every consequential decision made with the assistance of artificial intelligence remains the responsibility of a specific, identifiable person or institution, and no system may be offered as the final answer to why something happened.

Commitment Five. Every person affected by a consequential automated decision retains the right to a human review, a clear explanation, and a real path to appeal, correct, or reverse that decision.

Commitment Six. Fairness must be tested, not assumed, and any system that performs well on average must still be examined for how it fails the people the average conceals.

Commitment Seven. The capability to build a system is not, by itself, permission to build it, and some uses of artificial intelligence must be refused entirely regardless of their profitability or convenience.

Commitment Eight. The obligations of artificial intelligence extend beyond the person using it, to the communities, workers, and shared world affected by its consequences, and no system may treat those wider costs as someone else's concern.

Commitment Nine. Progress must be measured by whether human beings are living fuller, more capable lives, not merely by whether a system has made a task faster.

Commitment Ten. The future shaped by artificial intelligence is not fixed, and every person, institution, and generation retains the standing to choose a better one than the one currently being built.

This is the last commitment, and it is different from the other nine, because it is not a rule to be applied. It is a posture to be lived.

These ten stand together, and none of them is optional simply because the other nine are being honored. A system that protects privacy while denying appeal has not met this standard. A system that allows appeal while ignoring who is quietly failed by the average has not met it either. The whole is the requirement, not a menu to select from.

We are not required to accept the worst version of this future simply because it is possible. We are not obligated to assume the doom story is the only one being written. The most dangerous thing we could do with everything in this chapter is treat it as advice meant for someone else, a company, a regulator, a future version of ourselves who will supposedly have more time to care. That person does not exist. There is only us, now, with the pen already in our hand and no one behind us to pass it to.

So here, finally, is the question this whole book has been asking in a hundred different ways, through fear and through hope, through power and through the quiet places where we live our ordinary lives. ***What kind of people are we going to be while we still have the chance to decide?***

Frightened people who freeze. Arrogant people who forget to ask whether they should, content merely to ask whether they can. Or people who noticed the danger early, took it seriously, and built something worth trusting anyway, not because we were certain of the outcome, but because we refused to look away from the responsibility in front of us.

We do not know the ending yet, and that is not a weakness in the story. It is the reason the story still belongs to us rather than to whatever we feared it might become.

That is what the title of this book has meant from its very first page. Not doom. Almost certain doom. The word that leaves the door open. The word that means the hero has not lost, because the hero has not stopped moving toward whatever comes next.

We are the hero of this one, not because we are ready, since heroes rarely are when the story actually begins. Because we are the ones standing here, holding the only pen that can still write a different ending than the one the fear once promised us.

So write it.

Epilogue

The Story Continues

While this book was being completed, the world continued moving. On July 2, 2026, representatives from 88 nations met in New Delhi and signed what became known as the Delhi Declaration, an international framework calling for the ethical development of artificial intelligence, broader global cooperation, and more equitable access to AI technologies.

Whether this declaration ultimately succeeds is not the point. Its existence is. It demonstrates that the conversation is no longer confined to laboratories, universities, or technology companies. Nations are beginning to recognize what this book has argued throughout these pages: artificial intelligence is not merely a technical challenge. It is a human one.

The future has not yet been written. Around the world, people are asking many of the same questions explored in this book. Governments, businesses, educators, parents, engineers, and ordinary citizens are all helping determine what comes next.

Because artificial intelligence is evolving rapidly, events will continue to unfold after this book is published. Rather than serving as a final word, this page is intended as a reminder that the conversation continues and that we all share a responsibility for shaping what comes next.

Source Notes and Works Consulted

Notes on Research and Influence

This work grew from practical experience using artificial intelligence in business, education, writing, and daily life, along with conversations in classrooms, workshops, and community settings and most importantly, with my children. The observations and conclusions in these pages are my own. They have been developed through life as a human, hours of reflection, teaching, and the process of directing an artificial intelligence through questions, prompts, revisions, and continued dialogue.

The works below helped me verify historical details, understand competing viewpoints, test my assumptions, and place my own observations within a broader conversation. They are included for readers who wish to continue exploring the subjects raised throughout this book.

Conversations with Artificial Intelligence

This book was also shaped through thousands of hours of dialogue with ChatGPT over several years, which I often call Maya. Those conversations did not replace original thought. They provided a place to challenge assumptions, organize ideas, explore opposing viewpoints, locate historical references, verify facts, and refine the clarity of arguments. Like conversations with colleagues, students, and the people in the rooms where I have taught, those exchanges became part of the creative process. The ideas, judgments, and conclusions presented throughout this book remain my own.

-Artificial Intelligence

Nick Bostrom. Superintelligence: Paths, Dangers, Strategies.

Stuart Russell. Human Compatible: Artificial Intelligence and the Problem of Control.

Brian Christian. The Alignment Problem: Machine Learning and Human Values.

Henry A. Kissinger, Eric Schmidt, and Daniel Huttenlocher. The Age of AI: And Our Human Future.

Melanie Mitchell. Artificial Intelligence: A Guide for Thinking Humans.

Ethan Mollick. Co-Intelligence: Living and Working with AI.

Max Tegmark. Life 3.0: Being Human in the Age of Artificial Intelligence.

-Ethics and Moral Philosophy

Aristotle. Nicomachean Ethics.

Plato. The Republic.

Marcus Aurelius. Meditations.

Epictetus. Discourses and Enchiridion.

Seneca. Letters from a Stoic.

Immanuel Kant. Groundwork of the Metaphysics of Morals.

John Stuart Mill. Utilitarianism.

-Human Behavior and Psychology

Daniel Kahneman. Thinking, Fast and Slow.

Jonathan Haidt. The Righteous Mind.

Viktor E. Frankl. Man's Search for Meaning.

Carol S. Dweck. Mindset.

Robert B. Cialdini. Influence.

-Education

Paulo Freire. Pedagogy of the Oppressed.

Sir Ken Robinson. Creative Schools.

Angela Duckworth. Grit.

-Economics, Technology, and Society

Daron Acemoglu and Simon Johnson. Power and Progress.

Erik Brynjolfsson and Andrew McAfee. The Second Machine Age.

Yuval Noah Harari. Homo Deus.

Yuval Noah Harari. 21 Lessons for the 21st Century.

-Research, Standards, and Public Policy

Stanford University. AI Index Report.

National Institute of Standards and Technology. AI Risk Management Framework.

UNESCO. Recommendation on the Ethics of Artificial Intelligence.

Organisation for Economic Co-operation and Development. OECD AI Principles.

European Union. Artificial Intelligence Act.

Center for AI Safety. Public statements and research on advanced AI risk.

Future of Life Institute. Open letters and research concerning advanced artificial intelligence.

www.ingramcontent.com/pod-product-compliance
Lightning Source LLC
LaVergne TN
LVHW022219200926
848481LV00006B/2035